CW00729820

First published 2021 by the Tristan da Cunha Association

ISBN 978-1-7399910-0-5
A CIP catalogue record for this book is available from the British Library.

Layout and design by Direct Offset
Typefaces - Headings Minion Pro; Main text Adobe Garamond
Printed and bound by CPI Group (UK) Ltd, Croydon, CR0 4YY
Cover design by Murray Wallace (Direct Offset)

Nothing Can Stop Us

Tristan da Cunha's 1961 volcanic eruption and how its people handled their future

Tristan da Cunha from Nightingale Island (Antje Steinfurth)

Contents

Introduction

All places are unique, but Tristan da Cunha is extraordinary. An active volcano, the island is a haven for wildlife, including albatross and penguins. Its human history would be regarded as far-fetched if it were presented as fiction. The Tristan islanders are descended from an eclectic international mix of settlers whose arrival, survival, and the modern way of living they enjoy are all remarkable. The equality and communal living embraced by the community make it a special society, as they continue to live in the world's most remote settlement, Edinburgh of the Seven Seas.

All five northern islands showing Nightingale in the foreground, with its satellite islands of Middle or Alex and Stoltenhoff beyond, and on the horizon Inaccessible, left, and Tristan da Cunha, right. (Robin Repetto)

This book is published in 2021 to commemorate the 60th anniversary of the volcanic eruption on the island which led to the evacuation of the entire population to the UK in October 1961.

Its early parts set the physical scene of the islands before the coming of man and offer a fresh look at Tristan's history up to 1960. At the heart of the book is a detailed chronicle of the 1961-63 events covering the lead-up to that eruption, the island's evacuation, the refugees' time in England, and the episodes which led to the community's successful return. There are many twists and turns in this narrative, and the book continues by cataloguing the continuing challenges faced as the island's economy was re-built, leading up to a significant level of prosperity by the 1980s.

Twenty-first century Tristan da Cunha is explored in a different style, developing themes which include the threats and opportunities faced by the modern community. Finally, a reflective section poses important questions that Tristan is going to have to face in the future.

The authors have taken advantage of previously classified documents now available at the UK National Archives which has enabled them to explore how key people communicated behind the scenes during the unfolding events. Therefore, it is hoped the book provides a fresh insight into how provision for the refugees in the UK were made; evaluation of whether a return was practical; and deciding arrangements for repatriation after islanders declared that nothing could stop them going back. In this respect, and by the inclusion of previously unpublished personal reports and photographs, it is hoped the book will be welcomed by those already well acquainted with the island as well as providing a fascinating insight to those who read about this amazing story for the first time.

The book will also enable the reader to consider whether the decision of islanders to return to their isolated volcanic home was the right one, rather than opting to remain in a prosperous region in the southern part of England. Like the islanders themselves, who are all individuals, the resulting view will not be unanimous.

Finally, a word about the title we have chosen. 'Nothing can stop us' was a phrase used by Chief Willie Repetto when he was talking to the Danish travel writer Arne Falk-Rønne at Calshot, Hampshire in July 1962, and explaining to him with calm courtesy why the islanders were determined to return to their South Atlantic home. At that time there were many obstacles to overcome and a return was by no means certain, but we were both immediately touched by its aspirational and resolute tone.

Richard Grundy and Neil Robson 2021

Part 1: Setting the scene

Formation of the Tristan da Cunha islands - a hotspot indeed

Two hundred million years ago the continents we now call South America, Africa, India, Antarctica and Australia formed one massive supercontinent scientists have named Gondwanaland. The South Atlantic Ocean did not exist. But we live on a dynamic Earth, and about 190 million years ago a shift in the convection currents in the planet's core caused molten magma to well up between what were to become Africa and South America. Gondwanaland began to split, forming an opening that became a sea channel, and the continuing up-welling of magma along the mid-line of the rift fed into a new ocean floor, pushing the flanking continental masses apart. By 83 million years ago the ocean had extended northwards and widened to give us something like the South Atlantic we know today.

The main zone of up-welling and formation of the sea floor is still obvious as an underwater chain of volcanic mountains known as the Mid-Atlantic Ridge. The process continues today, and it causes numerous submarine earthquakes, but it was not responsible for building the Tristan da Cunha islands. These are the product of a so-called 'hotspot' of up-welling magma that began about 135 million years ago and created a string of volcanic cones, known as the Walvis Ridge stretching across the widening ocean from Walvis Bay in Namibia to the Mid-Atlantic Ridge. Tristan is the newest active volcano, although we cannot be quite sure that the other islands in the group are completely extinct. The other volcanoes on the Walvis Ridge have either been planed off by marine erosion to form flat-topped 'guyots' or are pyramidal volcanic cones known as seamounts that never grew to be islands.

Each of the Tristan da Cunha islands is at a different stage of evolution. Nightingale is the deeply eroded remnant of a large volcano whose oldest known rocks are about 18 million years old and now only 337 metres high as it has been much worn away, especially by the sea. Inaccessible, with rocks up to six million years old, is the eastern quadrant of what was once a nearly circular volcanic cone and was probably over 2500 metres high when active, but is now reduced to just 561 metres. Gough Island is formed from the same hotspot, despite being 370km away from Tristan. Gough was also built up around six million years ago, but there is evidence of relatively recent volcanic activity on all the smaller islands, and on Gough the summit cone may have erupted within the past 36,000 years.

The main island of Tristan da Cunha lies some 350km east of the Mid-Atlantic Ridge and began to grow when lava first erupted on the ocean floor 3500 metres below sea level about two million years ago. Thousands of separate eruptions emitted layers of ash and lava which breached sea level about one million years ago although the island may be younger. Further eruptions built up a cone over 2000 metres above sea level and over 4500 metres above the ocean floor. The huge volcanic pile is 56km in diameter on the seabed (12km at sea level) and estimated to contain 2,500 cubic km of volcanic rock. Lavas are composed of minerals including black pyroxene, grey feldspar, and green olivine, giving overall a bluish-grey appearance, locally known as bluestone. Volcanic ash, usually erupted violently in pyroclastic explosions, forms rocks which include a yellowy tuff – traditionally quarried by the Tristan islanders for building stone - and black rocks full of holes, like Aero chocolate, which float on water. Apart from being useful low-grade local building material, Tristan's rocks have no commercial value.

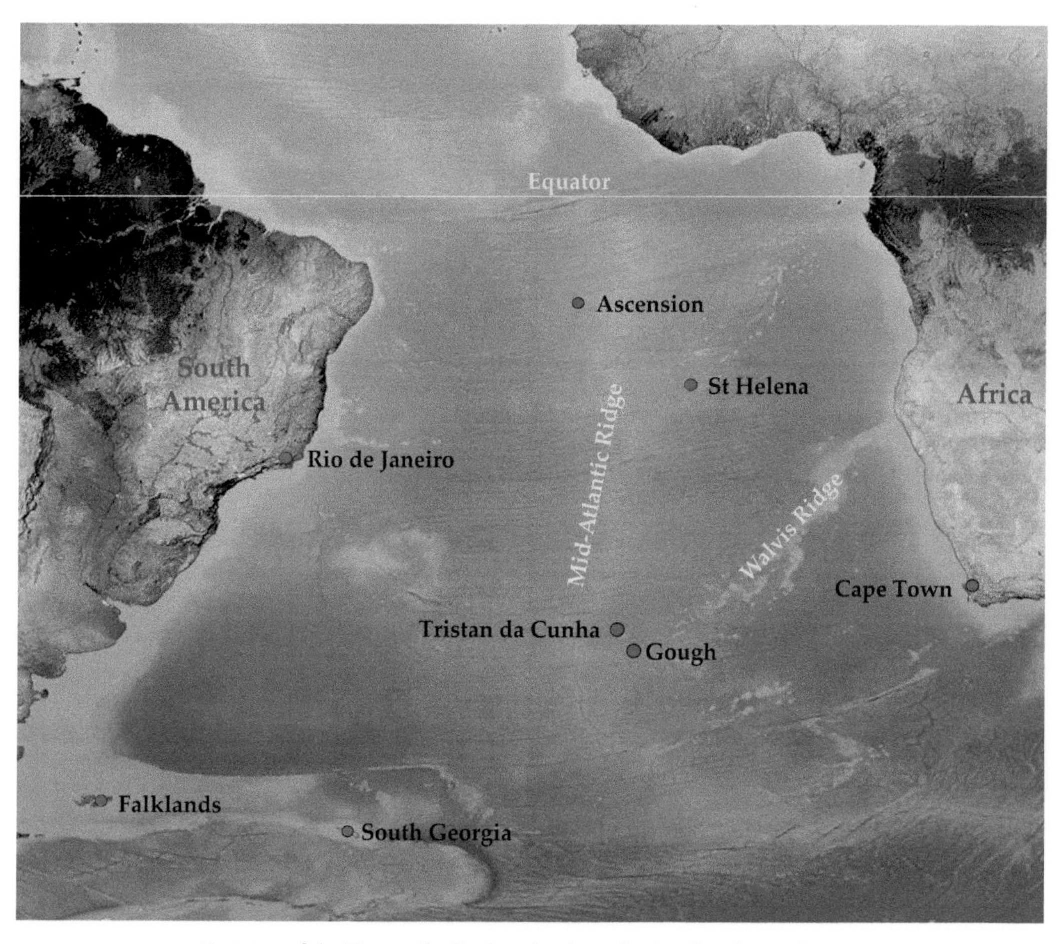

Position of the Tristan da Cunha islands in the South Atlantic Ocean.
(Based on a map sourced from NASA)

Distances			
Tristan da Cunha to:	**kilometres**	**miles**	**nautical miles**
Gough	**350**	**220**	**191**
St Helena	**2437**	**1514**	**1316**
South Georgia	**2666**	**1657**	**1440**
Cape Town	**2810**	**1750**	**1520**
Ascension	**3247**	**2018**	**1753**
Rio de Janeiro	**3344**	**2077**	**1806**
Falklands	**3886**	**2415**	**2098**

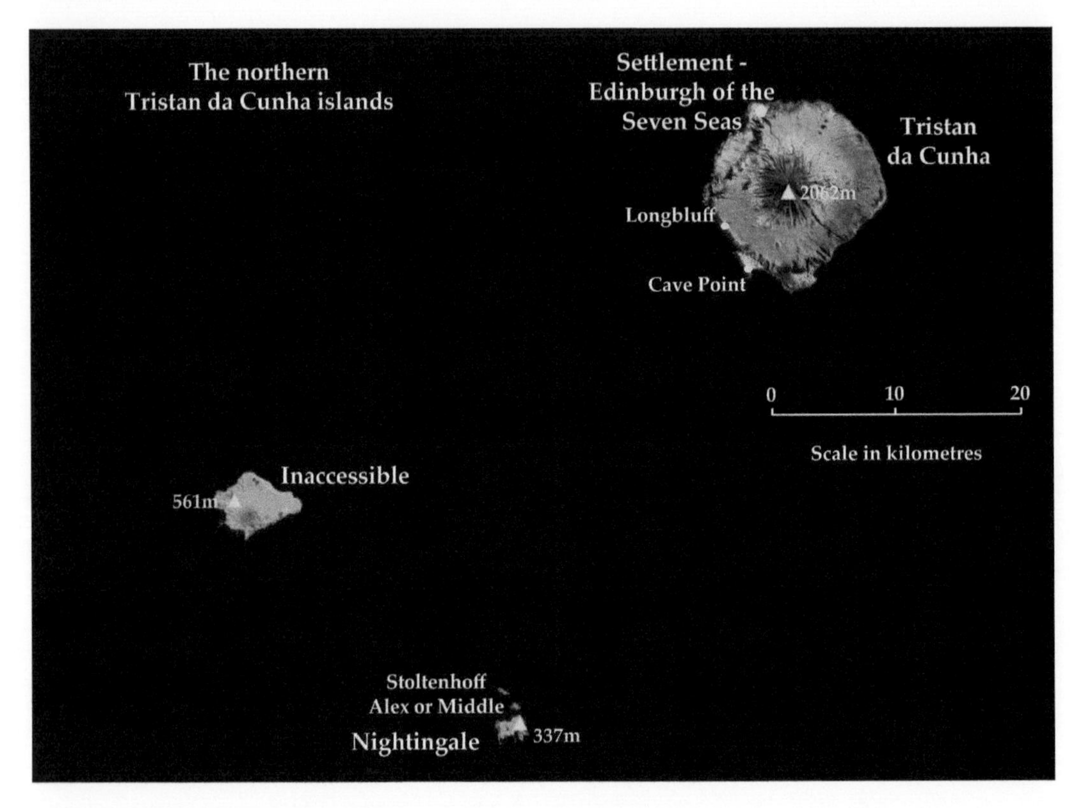

The Tristan da Cunha islands - see Section 7 for information on the Isolde Seamount.
(Based on a map sourced from NASA)

Distances			
	kilometres	miles	nautical miles
Tristan Harbour - Inaccessible	48	30	26
Longbluff - Inaccessible	40	25	22
Tristan Harbour - Nightingale	49	31	27
Cave Point - Nightingale	38	24	21
Inaccessible - Nightingale	22	14	12

The lava flows and deposits of tephra that erupted during the main shield-building phase formed sequences of rock now visible as distinct layers in the cliffs around Tristan. Many of these lava flows were originated from the summit crater or Peak, where the last eruption occurred about 5000 years ago. The volcano also has numerous dykes feeding magma to the flanks and giving rise to parasitic cones all around the mountain, as well as below sea level. One of these is Big Green Hill on the mountain's northern Base which is approximately 15,000 years old. A catastrophic collapse of the north-west side of the island resulted in one-fifth of the surface rock cascading into the sea and forming an enormous submarine delta fan. Behind the debris were exposed cliffs some 1300 metres high, by far the highest on the island. The collapse was probably a contributing factor which gave rise to the Hillpiece-Burntwood parasitic complex at sea-level 2,600 years ago which divided what became the Settlement Plain. In the south of the island a small parasitic cone erupted at Stonyhill about 250 years ago.

Big Green Hill parasitic cone on Tristan's northern Base (John Hodgkiss)

The Topography of Tristan da Cunha

Seen from an approaching ship, or examined on a map, Tristan main island looks a text-book example of a young volcano. It has a towering central cone – formally named Queen Mary's Peak but always simply 'the Peak' in island parlance - rising to 2062 metres above sea level. The highest point is part of the rim of a roughly circular crater about 400m across, with a small lake in its centre. The Peak is made of a series of lava flows alternating with layers of volcanic ash, some issuing from secondary centres rather that the main vent. The higher slopes are moderately steep and made of sparsely vegetated loose cinders but the angle eases lower down to around 8 degrees, forming a gently inclined, densely vegetated. plateau at between 600 and 900m known as 'the Base'. This is also broken by a number of secondary eruptive centres forming small, smooth, hills and more abrupt rock outcrops, while the three lake-containing craters known as 'the Ponds' were formed by explosive eruptions.

The Top Pond, one of three explosion crater lakes on the northern Base. Clothed mainly in fern bush and island trees, the grass slopes behind lead to Tristan's cinder Peak summit. (Paul Tyler)

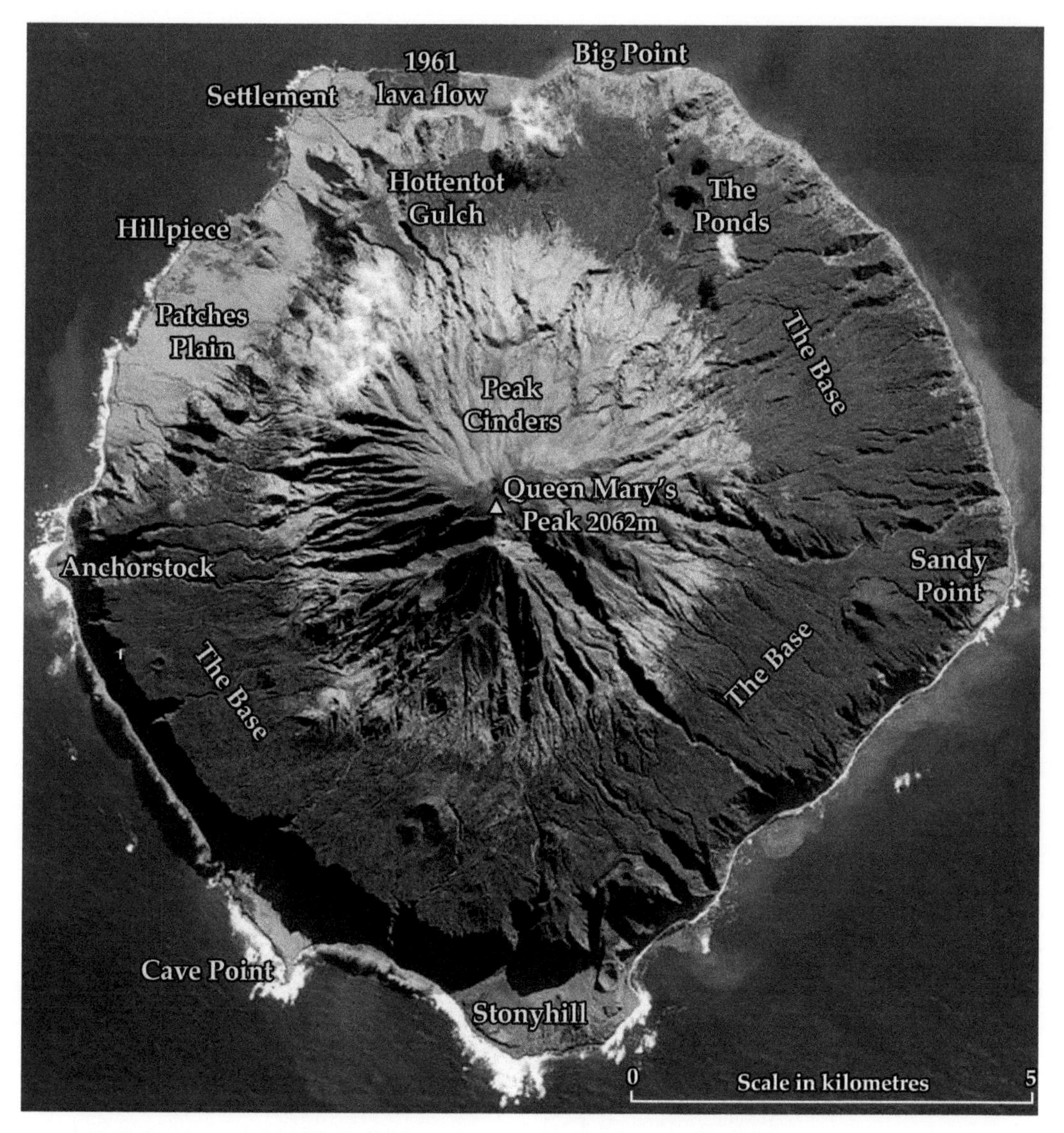

Annotated aerial image of Tristan da Cunha. (NASA)

Around much of the perimeter of Tristan the Base ends abruptly in cliffs and steep slopes, dropping to narrow boulder beaches, obviously the result of marine erosion. But on the north-west, east, and south of the island these cliffs are separated by areas of lowland at Sandy Point, the Settlement Plain, Cave Point and the Stonyhill plateau.

The island's climate

Tristan creates its own climate as saturated air is forced to rise over the mountain where it cools, forming cloud and heavy rain, or snow on higher slopes during winter. Rainfall exceeds 1600mm at sea level and more than double that on the high mountain slopes. Water readily percolates into highly permeable ash soils, but during storms heavy rain-wash soon removes coarse sand and rock particles. On the treeless upper slopes gullies have been cut, and lower down, where they break through the edge of the Base, they become deep gulches. During heavy rain, these gulches roar with the noisy tumbling of their load of volcanic debris, sweeping over the mighty sea cliffs, some more than 1000 metres high. The roar of a Tristan gulch waterfall, disgorging thousands of tonnes of boulders, can sound like thunder. But in dry conditions the only mountain streams are those fed from various peat bogs where moisture has been readily retained. However, ground water percolates easily/freely through the volcanic rock layers and appears as springs on the plains, feeding streams known as watrons, or it simply syphons up on beaches.

Tristan is situated 37° south of the equator, the same latitude as Melbourne and paralleled by Athens and San Francisco in the northern hemisphere. Unlike its continental counterparts, Tristan's climate is always moderated by the ocean, a factor that keeps mid-summer shade temperatures below 25°C but ensures that winters on the lowland plains are frost-free. The island creates its own micro-climate by always having a side in the lee of the wind. Air warms at it falls, thereby giving rise to calm, sunny and warm conditions even in mid-winter. When high pressure dominates, Tristan da Cunha enjoys glorious weather – cloud-free and mild in both summer and winter. On these occasions it is a beautiful place to be with crystal-clear air accentuating the vivid colours of land and sea.

Typical cloud formation caused by an uplift of saturated warm air as it is lifted and cooled as it passes over Tristan da Cunha (Team from RRS *Shackleton*)

Wildlife takes over

The emerging Tristan volcano was soon invaded by wildlife. The surrounding ocean teems with life, and an elaborate food web has a niche in the shallow waters and on the slopes of the growing volcano. Species already established on former and older islands quickly seized on the new environment.

Tristan rock lobsters under an overhang at Puma Rock with kelp attached to the seabed behind. (Sue Scott)

Seabirds and marine mammals like seals would have passed by as the volcano emerged, feeding on the life of the oceans, and they would have taken up breeding sites on the new land as soon as it had calmed and cooled. The southern right whales, once numerous around Tristan, would have begun to gather to breed in the coastal waters. The shallow seas would have provided a new habitat for seaweeds like kelp and a great diversity of other marine species. The marine flora and fauna of the Tristan group has close relationships with that of the coastlands of southern South America, aided by westerly ocean currents. Fish thrived and some evolved into unique sub-species in Tristan waters. Examples include the locally named soldier fish, bluefish, five finger and stumpnose. Larvae from the ancestor of the crawfish *Jasus tristani*, later sold as Tristan rock lobster, floated into local waters. They developed into scuttling crustaceans scavenging the seabed, despite the attention of local octopus, known on Tristan as catfish.

Tristan flora includes many species of moss, lichen and fern with small, light spores easily distributed by wind. Some plants have hooked seeds, able to hitch a lift on a bird. Spiders and insects are also readily airborne. Many of the islands' native invertebrates have closest relatives up-wind in temperate South America, indicating that this is where their ancestors almost certainly came from. But many kinds of invertebrates including centipedes, millipedes, dragonflies, ants, wasps and bees did not make it. Neither did frogs, toads, reptiles or land mammals until after people arrived.

Tristan's lowland plains would have supported a dense mass of *Spartina* tussock grass, while on the cliffs, in the gulches, and on the lower part of the Base 'fern-bush', dominated by the sole island tree, *Phylica arborea,* would have formed dense thickets. Above this zone the stubby dwarf tree-fern, *Blechnum palmiforme* would have dominated a band between 600 and 750m on the Base. Higher up, on the lower slopes of the Peak, there was probably a low-growing tussocky grassland with areas of the red-fruited crowberry or 'Peak berry', *Empetrum rubrum,* sedges, and mosses. Sparse crowberry, moss and lichen vegetation occurred on rocky outcrops higher up on the

Rockhoppers on Middle or Alex Island with Nightingale behind (Antje Steinfurth)

Mollies on the Tristan Base amongst bog ferns and island trees with Inaccessible Island behind. (Geoffroy Hienard)

peak. The bush and tree fern zones would have had a dense, spongy, peaty soil formed by the accumulation of sodden dead plant matter in the wet climate.

It was a bit easier for the land birds. Even today, when more than 3000km of ocean separate the islands from the South American continent, Tristan regularly receives frequent visits from egrets and the American purple gallinule known locally as the 'guttersnake'. Separate species of moorhen or 'island cock' did establish themselves on Gough and Tristan, as did *Atlantisia,* the elusive rail on Inaccessible Island. All three evolved flightlessness – the Inaccessible rail is the world's smallest flightless bird - presumably as they had plentiful ground food available. But other Tristan land birds can fly - including the charming Tristan thrush or starchy and the species of finch which evolved on all the main islands, including Gough, although the one on Tristan is now extinct. They are called 'canaries' by islanders, although not closely related to the familiar cage-birds.

The coastal tussock zone would have been the main breeding ground for a large population of Northern rockhopper penguins, and their adults still arrive in August and lay eggs in September. The young fledge by Christmas and the adults return in the new year to moult, losing their buoyancy so that they cannot swim, and adding an aromatic layer to the accumulation of guano while they wait for their new plumage. Great shearwaters would have nested in burrows in the peaty soil, as they still do on Nightingale, Inaccessible and Gough. As today, they would have made a huge migration northwards during the southern winter, circling the North Atlantic before returning to their southern breeding grounds in the following spring. Other ground-nesting petrels, including broad-billed prions or 'night birds', would also have been abundant. Subantarctic fur seals also bred in the tussock zone, where they were handily placed to hunt rockhoppers, while breeding elephant seals were formerly abundant on Tristan and a few still haul out on the sandy shores.

Peeoo displaying above Salt Beach, Inaccessible Island (Peter Ryan)

Three albatross species breed on the Tristan Islands. The smaller Atlantic yellow-nosed albatross or molly *Diomedea chlororhynchos* still builds its pedestal soil nests on the mountain Base, laying eggs in September. They hatch by November and by Christmas the chicks can be left unattended while both adults forage for food. Chicks fledge ready for flight by April. The other small sooty albatross chooses cliff ledges for protection and easy take-off. Tristan islanders heard their haunting calls and consequently decided to give the bird the onomatopoeic local name of 'peeoo'.

Mollies and peeoos are still numerous on Tristan, where they are now protected, but the greatest of the albatrosses, the Tristan albatross or 'gony', now breeds only on Gough (apart from two or three pairs on Inaccessible). But before the advent of humans many gonies bred on the lower slopes of the Peak of Tristan, on open windy slopes where they could take off and land safely, although needing a long run to take to the air in calm weather. Gonies have a wingspan of over 3 metres and are a majestic sight as they soar effortlessly above the ocean waves, though they are quite ungainly on the ground. Having returned to nests in November and laid their eggs by January, gony chicks are protected throughout the winter months, staying to fledge and departing the following mid-summer between December and February.

So, Tristan's original wilderness supported a special blend of plants, seabirds, seals and penguins, surrounded by bio-rich seas. Many kinds of plant and animal familiar on the continents would have been absent, excluded by the vast ocean barriers all round, for the island and its neighbours are in extreme isolation, lying 2160km from St Helena, 2800km from southern Africa, and 3200km from South America. In that isolation, evolution has given rise to a number of species found nowhere else in the world. Tristan is an active volcano with a wet and windy climate where narrow shelves of densely vegetated lowlands once flanked towering scrub- and fern-clad cliffs and, above them, harsh mountain slopes. Surely, this is a place for only species like penguins and albatross to occupy? Yet this extraordinary spot has become a corner of the world where people have freely chosen to live, and where they survive against all the odds.

A Wilkins finch flying from a frond of island tree on Nightingale Island (Ben Dilley)

Part 2: A Fresh Look at the History of Tristan da Cunha

From discovery to settlement

Within the vast collection of western manuscripts in the British Library there rests a folio of great interest. It contains a number of detailed seventeenth-century accounts of significant Portuguese naval expeditions. Search amongst the many pages of this Relação du Naos e Armadas da India and the following remark comes to light:

> In this voyage, Tristão da Cunha discovered the islands to which he gave his name, and thus they are called at the present day.

The Portuguese explorer Tristão da Cunha (c.1460-c.1540), an engraving from about 1840 by João Macphail. (National Library of Portugal)

What a throwaway line. Yet this simple statement, presented here in a Victorian translation, marks the start of the chain of events leading to the foundation of the quite remarkable society that forms the subject of this book. It refers to an eminent sixteenth-century admiral who set forth with a fleet of fifteen ships on a mission to assert his nation's dominion over certain Arab cities on the east coast of Africa. His name, of course, was Tristão da Cunha and the year in question was 1506. During his journey he chanced to catch a fleeting glimpse of the Tristan group though, sad to say, there is no evidence that any of his ships' men were able to land on the islands; all da Cunha could do was pass them by.

In the 150 years that followed many more explorers, both navigators and scientists, visited the archipelago, a number of them actually getting ashore but none of them making any particular long-term impact. The first man known to have squatted there for anything like a reasonable period was a seal hunter from Philadelphia named John Patten, the master of the *Industry*, who eked out an existence on the main island with a number of his crew between August 1790 and April 1791. They pitched their tents beside its north-west shore on the site of the present settlement, and during the course of those nine months they secured some 5600 seal skins for onward transfer to the lucrative China market.

But the first true settler with any intention of staying permanently was the American adventurer Jonathan Lambert. Arriving in January 1811 with a couple of associates, he settled in the north-western corner of Tristan on the site of the present settlement. Goats and pigs had already been released on the island, and he brought with him not only more pigs but also some geese. He and his companions cleared and enclosed about 5 hectares of land and grew potatoes, cabbages, maize, radishes and pumpkin vines. While their crops and livestock developed, the party subsisted largely on elephant seals which they also fed to their pigs, and in addition they also ate the endemic moorhen which they found 'fat and delicate'. There was an arrogance if not a touch of romance to him for, on 4th February that year, he drew up a proclamation declaring that he had 'taken absolute possession' of the islands which 'shall, for the future, be denominated the Islands of Refreshment'. 'We suppose [this term] to indicate the indulgences with which strangers will meet upon visiting *his Majesty*,' joked *The Observer* newspaper in London at the time. His intentions, not fulfilled, were twofold: to establish Tristan as a victualling-place for ships passing on what was a main sea-route between Europe and the Indies, and to render large quantities of elephant-seal oil and store the stock he produced in stone-lined cisterns for later sale in Rio de Janeiro. But his enterprise soon failed and by early 1812 he was pleading with the Cape Government to take charge of his realm in return for a monthly income. Not that he lived much longer to enjoy his sovereign existence, since he disappeared with two companions on a fishing trip in May 1812.

In the years that followed, the strategic value of Tristan gained strength from two contrasting sets of circumstance. In Europe events of epic proportions were gathering momentum. After a long struggle Wellington routed Napoleon's forces in the war on the Iberian Peninsula, leading eventually to the French dictator's capitulation and exile on the island of Elba. His subsequent escape and the touch-and-go victory at Waterloo resulted in his banishment to St Helena. This saga of affairs resulted in such a level of paranoia that it was feared that Bonaparte's partisans might use Tristan as a springboard from which to bring about his release, though with hindsight the thought that such an enterprise might ever have been successful beggars belief. Rather more significant were the lingering transatlantic tensions accompanying the so-called 'War of 1812' between Britain and the United States. With enemy cruisers and privateers ominously using Tristan as a point of rendezvous, and its favourable situation proving ideal for supporting foreign whaling activities, Britain's Secretary of State for War and the Colonies Lord Bathurst felt eventually prompted to despatch the frigate *Falmouth* under the command of Capt. Festing to establish a modest military presence on the island. Festing took with him seventeen junior naval officers and men and, when they landed on 14th August 1816, they discovered a couple of stragglers who reportedly 'were both overjoyed in placing themselves under the protection of the British flag'. One of these individuals was Thomas Currie, a native of Livorno and a remnant of Lambert's party, who – if legend is to be believed – had accumulated a considerable fortune in the form of gold coins from selling seal skins, oil and provisions to passing seafarers. These riches, it is said, he buried, and they are yet to be rediscovered.

In November 1816, the detachment was relieved by an army officer, Capt. Cloete, accompanied by a mixed company of about forty men and their families and several aboriginal South Africans. But, as *The Times* later put it, 'when all fear of Boney was over,' the sheer logistics of keeping the garrison supplied with provisions from the Cape proved too costly, and Cloete and most of his party were withdrawn in May 1817. What followed came to constitute the island's first major tragedy. Late that September the Cape authorities despatched HMS *Julia* to pick up the people left behind, but during the night of 2nd October a heavy swell smashed the sloop against a reef and 55 of the crew were drowned. As a result, it was only in November 1817 that final efforts were made to vacate the settlement, though one man out of Capt. Cloete's company, William Glass, begged leave to remain along with his wife and two children.

Born in 1787 Glass came from the Scottish Borders. He was a corporal in the Royal Artillery, a man of strong religious faith, and a member of a Freemasonry lodge at the Cape. Two others also agreed to stay, a pair of stonemasons from Devon, Samuel Burnell and John Nankivel, and right from the start Glass's high moral influence was crucial in defining the culture of this fledgling society. Most notably, he was instrumental in drafting a document whose importance it is hard to overestimate. In it he refers to the three of them – just the men, that is, not his wife – as a 'co-partnership', and defines the terms of their voluntary agreement like this: that all possessions 'shall be considered as belonging equally to each', that any profits they make 'shall be equally divided', and that to maintain harmony 'no member shall assume any superiority whatever'. In contrast to the deep class-divisions that characterised British society before the passing of the Reform Act of 1832 this agreement was radical stuff, and it was duly witnessed on 7th November 1817 by the garrison's remaining senior officer shortly before his departure. Nankivel's signature took the form of a simple cross. At last the foundation of this unique community had been achieved, and the basic principles of that declaration may still be found at the heart of the people of Tristan right down to the present day.

The early settlers' cry: 'We need more women!'

From the beginning Glass was faced with two formidable questions: how could this newly founded settlement evolve into an economically viable society? And how could a virtually single-sex community ever hope to survive for any length of time? In the case of the former, the settlers' stockholding was a meagre one: just six horses, four cattle, fifteen sheep, about forty pigs and some ducks and turkeys, together with a feral population of other beasts and a modest parcel of land cleared for grazing and the cultivation of crops. At least elephant-seal oil and sealskins would ensure a reasonable level of income through trade with the Cape as well as providing commodities for bartering with passing ships. But the only female on the island was Glass's wife Maria, a South African Creole who was destined over the following four years not to see another woman.

Maybe she occasionally reflected on this deficiency in her life. But an irony awaited her because, when the opportunity finally arose, the experience proved to be a ghastly nightmare. On 23rd July 1821 in a heavy swell and surrounded by mist the East Indi-

The cluster of cottages forming the early settlement on Tristan, a watercolour by Augustus Earle in 1824. (National Library of Australia)

aman *Blenden Hall* was driven towards the breakers around Inaccessible Island and smashed to pieces. Two men were drowned but something like 52 passengers and crew – accounts vary – made it onto that aptly-named island's forbidding beach. Many of the survivors landed literally naked, and all were reliant on the fragments of wreckage slowly being washed ashore. They managed to erect a few tents using some bales of coarse cloth which had floated towards them, but almost immediately disorder took hold. The crew turned mutinous, while the women divided themselves by class – at least two amongst them viewing themselves as 'ladies' – and the quarrels that ensued were incessant, all of which made the situation deplorable for the castaways over the following three-and-a-half months. It was only after more than one attempt that two of their number in a flimsily patched-together boat managed to get across to Tristan to raise the alarm. The survivors were transferred in batches to the larger island, where they were generously made welcome. But not for long. The womenfolk continued to bicker indefatigably, and the crew descended into such a state of insubordination that, to the distress of the islanders, a number of them were flogged on their captain's orders. Glass struggled on helplessly while this mob of ill-matched individuals reduced his island to a state of near starvation and ruined its precious equanimity.

It was not until 9th January the following year that a passing brig took them off to safety at the Cape. The relief on Tristan is easy to imagine, added to which the cloud had a silver lining. Two of the survivors – Stephen White, one of the crew who had been flogged, and a young woman called Peggy, the long-suffering maid of one of the 'ladies' – elected to stay on the island, thereby increasing its population to eleven men, two females and an expanding number of Glass and White children. As William Glass said to a passing sailor on the *Berwick* in March 1823, 'if [we] had but a few women more, the place would be an earthly paradise.'

Over the following few years, a certain Capt. Simon Amm, an otherwise minor character in the story of Tristan da Cunha, made a quite disproportionate contribution to its development. In February 1824 he left Rio in his sloop *Duke of Gloucester* bound for Cape Town with the talented artist Augustus Earle among his passengers. During the following month, despite the wind blowing ferociously hard, Earle ventured onto Tristan with just his dog, a gun and his painting materials, whilst the ship took on a supply of potatoes. By the following morning the storm had turned into a gale, and he could only watch in horror as Amm tacked his vessel out to sea and disappeared. Suddenly he found himself in a most distressing situation, and as he wrote in a letter to the *Hobart Town Gazette* in February 1825, 'eight dreary months did I endure on this dismal and sequestered spot, in a state of anxiety and expectation indescribable.'

There were, however, a number of diversions to help him struggle against these gloomy thoughts. He was charmed by the settlement's incongruous cluster of six little houses with a Union Jack fearlessly fluttering over them, and he came to feel a sincere respect for Glass, a man he accurately recognised as kind, hardworking and devout. But he found it difficult to engage with some of the others in the community, rough British tars whom he later described as lacking 'much of elegance or refinement, or [any] conversation … such as would be tolerated in polished society'. For all this he was temporarily appointed chaplain and schoolmaster and, until his painting materials ran out, he found consolation in his love of art. Indeed, many of his fascinating watercolours survive and may still be seen in the National Library of Australia in Canberra.

Half-a-dozen times ships passed tantalisingly close to the island without pausing and it was only with the arrival of the *Admiral Cockburn* on 29th November that this unwilling visitor was able to leave and travel on to Van Diemen's Land, now Tasmania. By 1832, however, when he wrote an account of his adventurous voyage, he had recovered enough from his traumatic experience to almost forgive Simon Amm and to place on record his 'sincere gratitude to Glass and his companions for their unremitting kindness notwithstanding all the trouble I had given them'.

Did Capt. Amm feel any guilt over leaving Earle to his fate? Some have suggested he may have done. Certainly, he returned to Tristan a number of times, notably during 1826 when, in response to a plea to seek out more settlers from the Cape, he brought Thomas Swain, a 52-year-old Sussex-born seaman who had endured the uncomfortable experience of being captured during the Napoleonic wars as a British deserter by the French, agreeing to fight on their side, and then being captured once more, only this time by his native countrymen. Swain became a permanent addition to the Tristan community, and his name is still found on the island today. But Swain's arrival hardly helped its gender balance, so Amm was charged with the task of inducing five St Helenian women who were searching for husbands to come and live on Tristan. On 12th April 1827 he arrived back with his cluster of passengers, most of them women of colour, and possibly a couple of them ex-slaves. In terms of character they may not have been the most appropriate candidates for the venture, but the alternatives available to them were probably limited, and in any case, within the shameful values of the time, they must have known that they were a negotiable commodity. They undeniably satisfied a pressing requirement, while Amm, according to tradition, received 20 bushels of potatoes in payment for each of them. Thomas Swain had already determined that he would marry the first of the women to land on the beach, and that honour fell to Sarah Jacobs. They promptly became a couple. Four other men followed suit.

This was a time when sailing vessels from the eastern coast of America called evermore frequently, intent on hunting the Southern right whale, an animal apparently named that way because it was the 'right' ones to chase, being easy to catch because it moved slowly and floated when dead, so that its blubber and whalebone could be cut while it was tied alongside a whale-ship. Naturally, the crews of these vessels were eager to barter staple foods, clothing and other necessities in return for fresh water and meat, and for a while the community was totally reliant on this trade. Shipwrecks added further settlers too, the most significant being the foundering of the *Emily* in September 1836. One of the men who struggled ashore was a Dutchman in his late twenties called Pieter Groen. Born in the coastal town of Katwijk in the province of South Holland, he would go on to anglicise his name and marry one of the daughters that Sarah Jacobs had brought with her. Another settler in 1836 was the American whaler Thomas Rogers, followed in 1849 by his fellow compatriot Capt. Andrew Hagan, a seafarer so disenchanted with the life he was leading that he jumped his own ship. Each of the men married one of the numerous daughters produced by William and Maria Glass. By 1840 the number of separate families' resident on the island had grown to nine; by the 1860s 'Peter Green' had become the community's leading man.

In April 1843 amidst its tightly printed columns *The Times* carried a revealing account of Tristan's situation from a traveller on his way to Madras, now Chennai. In it he reported that the islanders' assets were fifty head of cattle and 100 sheep, 12 acres (about 5 acres) of potatoes and plenty of apples and peaches. Their most pressing want, he noted, was a stock of nails to repair their houses recently damaged by fierce winds. Not for nothing did the heading to the article carry the title 'A singular community'.

Prosperity begins to dim

For many years God-fearing William Glass had recognised the settlement's need for education and spiritual leadership, and this feeling he no doubt made with some force to Rev. John Wise, a minister travelling to Ceylon, now Sri Lanka, on the Augusta Jessie in October 1848. Wise went ashore a number of times, and he subsequently reported his observations to the Society for the Propagation of the Gospel (SPG) in London who went on to publish his plea in the monthly press. By 1850 an anonymous benefactor had donated the considerable sum of £1000 to provide the island with a resident clergyman, and as an upshot of this request Tristan's first missionary arrived on the island in February 1851. His name was Rev. William Taylor.

Three years later William Glass died of a particularly invasive form of cancer at the age of 67, and the community lost not only its patriarch but also, with their departure in January 1856, the entire Glass clan who went to join various relatives in New London, Connecticut. Thereupon the Glass name disappeared from the island until Thomas Glass returned ten years later. Taylor's initial report to the SPG was positive, but he became increasingly aware that American whalers were calling less often, and with the islanders reliant on these visits for flour and other commodities, his fears intensified for their future. He therefore advocated a total evacuation of the community, and at the urging of the Bishop of Cape Town, some forty-five islanders accompanied Taylor aboard the paddle sloop HMS *Geyser* to settle in his new parish at Riversdale in the Cape in March 1857. Only four families remained, just twenty-eight people, the men with arguably the most influence being Peter Green and Thomas Swain. A pause for thought indeed.

Up till then the island had been able to maintain a reasonable trade with passing whalers and sealers, but with mineral oil taking the place of its animal equivalent in everyday usage that traffic essentially evaporated. The American Civil War made an impact too. The Confederate commerce raider *Shenandoah* had been instructed to destroy the northern states' whaling fleet. Passing by Tristan during 1864 the ship's commander demanded proof that Tristan was indeed a British possession. The islanders were unable to provide any – indeed, no such document had ever formally existed – whereupon the officer proceeded to land around forty Union prisoners without any provision for their keep. Luckily, they remained for only a few days before they were picked up by a gunboat belonging to their own side.

Oswald Brierly's sketch of the scene on the day of the first Duke of Edinburgh's visit with Peter Green's house to the right. (*Illustrated London News*)

Under more agreeable circumstances, in August 1867 Queen Victoria's second son, the Duke of Edinburgh, became the first British royal to set foot on this lonely outpost of empire. Aged just 22 he had recently been appointed commander of HMS *Galatea*, a steam-and-sail frigate with 26 guns and a sizeable complement of officers and men. Having made a turbulent crossing to shore in a naval cutter the duke, together with a party of officers and officials, toured the eleven stone cottages which formed the remote little settlement, and spoke to all of its 53 inhabitants in succession, among them Green's wife, a woman whom the ship's chaplain subsequently described in spirited fashion as 'buxom [and] merry-looking'. There followed a dinner comprising roast mutton and poultry with potatoes and parsnips, but the only liquid refreshment was 'the purest water imaginable' apart from a bottle of wine that proved to be practically undrinkable. Finally, Green accompanied the visitor back to the *Galatea* and before returning to shore he diffidently asked the duke's permission to call 'their little village' Edinburgh. This was agreed, and so it has been known ever since.

Benjamin Shephard's sketch of HMS *Challenger* off Tristan da Cunha on 14th October 1873. (Philadelphia Maritime Museum)

The harrowing isolation that came to define the perilous existence of the islanders well into the twentieth century was yet to take hold, but clearly life had to change. Steam was replacing sail as the principal mode of propulsion, whilst the opening of the Suez Canal in 1869 meant that the traffic following the Atlantic sea route from Europe to India and Australia, often calling at Tristan along the way, steadily dwindled. Trading would have to give way to subsistence farming and fishing. Scientists, nevertheless, were attracted to the archipelago, a prominent example being the visit in October 1873 of HMS *Challenger*. Its round-the-world expedition was a voyage of significant scientific discovery since it laid the foundations for making oceanography a research discipline. During the *Challenger*'s four-day stay the two islands of Nightingale and Inaccessible were systematically surveyed, whilst the zoological discoveries in the surrounding waters proved to be legion. The expedition members were fascinated by the sociological aspects of the small community on Tristan itself, its appeal captured in this lively if ungrammatical note of theirs:

> The colonists live in houses built of large basaltic blocks, shaped with the axe and made to fit each other closely, as there exists no lime on the island to make mortar with.

They even experienced a touch of drama. While on Inaccessible the crew were able to rescue a pair of German brothers who had settled there, subsisting by whatever means might come their way, and leading for two years, in the words of the subsequent report, 'a sort of Robinson Crusoe life'.

The late 19th century sees catastrophe and renewal

The year 1881 saw the arrival of Tristan's second chaplain, Rev. Edwin Dodgson, a man worthy of note if only because he was the brother of Lewis Carroll, the author of *Alice in Wonderland*. By then the community had grown to over 110 people, though cruel reversal awaited it over the next few years in the form of two appalling disasters. First, in 1882 an American schooner, the *Henry B. Paul*, was wrecked on the east coast of the island at Sandy Point. The rats infesting its holds jumped ashore, and within three years the rodents had spread across the entire landmass, destroying a large portion of the native birdlife as well as the islanders' crops. The Tristan men, reduced by desperation to trading with any passing ship whatsoever, began to take increasingly perilous risks to barter for precious flour, sugar, clothes and other luxuries such as soap to supplement their impoverished lifestyle. Weakened by his three years on the island Dodgson left Tristan in 1884, reporting to the SPG that the long-term future for its population was unsustainable, what with its extreme isolation and its reliance on a fragile balance of agriculture, fishing, hunting sea birds and gathering their eggs. For him the removal of the entire population was the only solution.

A far greater tragedy followed when on 28th November 1885 fifteen of Tristan's eighteen able-bodied men rowed out in a lifeboat in an attempt to trade with the barque *West Riding*. Struggling through the waves on their quest every one of the crew was lost. At a stroke Tristan was rendered what Peter Green described as 'an island of widows'. One person shocked by the tragedy was Rev. Dodgson, by then back in Britain, and he made immediate plans to return to help. Another man reading of the calamity in London the following March was a 23-year-old clerk, Douglas Gane. In July 1884, while travelling to Australia, he had actually seen the island and met some of its menfolk when the clipper *Ellora* in which he was travelling anchored in a heavy swell some 8 miles (12km) off Tristan. As Gane's son Irving reflected at the time of his father's death, the news of the lifeboat disaster 'must have made a deep impression' on him. True indeed! The value of Douglas Gane's subsequent advocacy on behalf of the island would prove incalculable.

On occasion, the fates were kinder. The shipwreck in October 1892 of the barque *Italia*, which caught fire and was run aground at Stony Beach on the southern side of the island, provided a pair of much-needed male immigrants. The ship's petty officer Andrea Repetto and his shipmate Gaetano Lavarello seized the opportunity to settle down, each marrying a local woman and adding his skills to the tiny community. Andrea, intelligent and confident, soon took a prominent role in Tristan's affairs, whilst Gaetano pioneered the local construction of longboats, their light wooden frames covered with stretched canvas proving ideal for hauling up onto rocky beaches and, as seabirds and seals became increasingly scarce on Tristan, perfect for hunting and gathering trips to Inaccessible and Nightingale islands from where eggs, meat, and cooking fat could readily be obtained.

A striking sketch of Peter Green from an 1897 issue of the *Graphic*, a British weekly much admired for its high-quality illustrations. (Westminster Libraries)

By the final decade of the nineteenth century Peter Green, an elderly man but still leader of the community, had become widely credited with having saved literally hundreds of lives from shipwreck, and his actions, combined with those of the other islanders, attracted recognitions of the highest order, among them a gold watch from President Rutherford Hayes of the United States following the sinking of the American barque *Mabel Clark* in 1878, and a silver medallion from Umberto I of Italy in 1892. But beyond doubt the greatest honour he was to receive came from Britain's Queen Victoria in the form of a huge facsimile of a portrait of herself by the Austrian artist Heinrich von Angeli, set in a gilt frame surmounted by a small Imperial crown and signed by the monarch herself. Initially despatched to St Helena, the fragile package arrived aboard HMS *Magpie* in November 1896. Green was thrilled when he saw it. 'Such a picture never came to Tristan before,' he gasped, though for all that the crate was specifically addressed to him, it was surely an acknowledgement of the debt of gratitude owed to the island as a whole. Nor can its political dimension be overlooked; it may have been the benign image of the head of the Empire but it was also an emblem conveyed across the ocean of the power and protection that Britain wished to bestow on its most isolated possession. Peter Green died aged 94 in 1902 and yet his gift survives, hanging in a proud position in the community's Anglican church of St Mary. The frame may be chipped, but Victoria's signature - still faintly visible - glimmers hauntingly in its bottom right-hand corner.

The British Government loses interest

At the start of the last century, the island could justifiably expect a call once a year from a Royal Navy ship, a routine established with the visit of the naval corvette Thalia in 1886 and lasting right up to the arrival of HMS *Odin* in January 1904. Significantly, its captain brought with him an offer of free passage to South Africa as an inducement to leave Tristan, accompanied by a promise of plots of land for the islanders to cultivate at the Cape. There were no takers. By then, however, the British Government felt reluctant for reasons of economy to continue these annual visits, despite the fact that the despatch of a warship from Simon's Town cost no more than £400.

Undeniably the island had its uses as a place of call for trading vessels, but from now on the islanders would be dependent on passing ships' captains for news of the outside world. As a resourceful community it tried to maximise its potential for agriculture but it was hampered by large stock numbers, estimated at 700 cattle and 800 sheep in 1905, which meant the Settlement Plain's grassland became over-grazed and its animals under-nourished. However, the wet and windy winter of 1906 saw the death by starvation of some 400 cattle, their loss paradoxically giving the survivors a greater share of meagre pastures.

What a strange community these visiting crews would have found. Here was a micro-society of about eighty inhabitants clad in moccasin-style shoes of bullock hide and home-made socks knitted out of pure wool. It followed almost communistic habits under which pastureland was held in common. Its members drove primitive oxcarts and, according to a letter in the *Illustrated London News* from Lieutenant Traill Smith of the *Odin*, they spoke English 'very slowly like Devonshire drawl'. In the words of Andrea Repetto, 'The wind blights our plantations, so we are very short of potatoes,' and with the frequent lack of tea, coffee and sugar it was indeed a hard life.

Island men assembled in 1906.
Top row: Sam Swain, Bill Green, Andrew Swain, Andrew Hagan, Willy Swain, William Rogers; middle row: Old Sam Swain, Gaetano Lavarello, Henry Green, Bill Rogers, Bob Green, Andrea Repetto; front row: Ben Swain, John Glass, Alfred Green, Charley Green, Tom Rogers.
(Tristan Photo Portfolio)

One such vessel to visit this desolate spot in 1905 triggered disproportionate publicity for the islanders. This was the steam yacht *Pandora* under the captaincy of the entrepreneur Thomas Caradoc Kerry. He had been granted a Government concession to remove guano from Inaccessible and Nightingale islands for commercial purposes.

Exasperated by the stack of gifts cluttering his decks from well-wishers, he allegedly threw many of the parcels overboard, notably a consignment of Bibles from the SPG, which he was said to have described as 'a load of old rubbish'. Back in Britain there followed a trial for theft which was eventually halted on the grounds that it had been unreasonable to ask Kerry to carry '1500 books for a few uneducated islanders'. Perversely, the scandal delivered the benefit of exposing the serious privations within the Tristan community at the time.

In April 1906 Rev. Graham Barrow, a clergyman from Malvern, responding to a plea from Tristan to the SPG, came out as a resident priest and teacher, accompanied by his wife Katherine and, amazingly, a maidservant. Barrow had an unusual incentive. In 1821 his mother, then a little girl of 4, had been among the survivors from the *Blenden Hall* disaster; so here was his way of thanksgiving. The Barrows stayed until 1909. At one point during their tour of duty, the South African ship owner Casper Keytel sailed to Tristan at the insistence of the Colonial Office to investigate trading opportunities with the island. In March 1908 he made a second trip to Tristan with a group of former islanders including three men who had married three sisters Annie, Elizabeth and Agnes Smith. The Smith girls, all Irish, were brought up in the Roman Catholic faith, and, while they were content to join Anglican services, it was Agnes who in later years would lead a small breakaway congregation which resulted in the current thriving Tristan Catholic community.

Otherwise, these lean times continued without a break and the public's perception of the island community became so detached that a contemporary Blue Book, one of those regular official summaries of colonial data, allowed itself this sardonic remark:

> Tristan da Cunha is generally supposed to be inhabited by a dusky British race who live on fish, the spoils of the wrecks which strew its iron-bound coasts, and the publications of the Religious Tract Society.

World War One's heavy hand stretches far

The outbreak of World War One only deepened the lack of shipping. In August 1916, amid its grim descriptions of the horrors of battle on the Somme, *The Times* reported that a whaler from Norway would be calling at Southampton to pick up mails for Tristan and appealed for clothing, groceries and sewing materials for its inhabitants. The following month the same paper reprinted an account in the New Zealand press from the captain of a French sailing ship *Bonneveine* which had recently called at the island.

On arrival 'the first cry of the inhabitants was for flour,' he reported. They were clad in skins and mere remnants of civilised clothing 'and some, indeed, wore nothing'. The community had practically gone feral. This news acted as a catalyst for Douglas Gane, who was by then a successful solicitor in the City of London, and within a day he added his weight to the appeal for staples and clothing. Thus began his determined efforts to badger the authorities into showing more respect for the settlement and its needs. Finally, in July 1919 he was successful in persuading the British Government to send the light-cruiser *Yarmouth* laden with stores, mail, and confirmation that the First World War had come to an end. In 1920 Gane received a letter from an unidentified Tristanian mourning all those who had died in the conflict and adding a hope that 'perhaps some day a clergyman may come to live amongst us'. Gane had a focus at last, and in February 1921 he issued an appeal for subscriptions to a 'Tristan da Cunha Fund', ostensibly to support the sending of a minister-cum-schoolmaster on a short-term posting under the supervision, as usual, of the SPG.

So exceptional was the position that it was far from easy finding an appropriate candidate. Eventually a Leicestershire curate, Martyn Rogers, was chosen though it was not until early 1922 that a suitable passage was found for him and his wife, Rose. They were warmly welcomed by the island's inhabitants keen for leadership and education. Within two months of their arrival the polar expedition vessel *Quest*, returning to Britain following the death at South Georgia of the explorer Ernest Shackleton, stopped at Tristan to carry out scientific work. All did not run smoothly, however, and its commander, Frank Wild, was appalled by the unruliness of the men as they ran amok around the decks. They 'crowded aboard in dozens,' he later recalled. 'Immediately there was a noise like babel let loose' as they badgered the crew 'in thin jabbering voices' for everything they could spare. They were 'an uncouth lot,' he grumbled, proving such a nuisance that he felt obliged to order them off his ship, though he subsequently reflected on his brusque reaction, adding that he believed 'the older men among them were really grateful for what we had been able to do'.

Also, on the *Quest* was the expedition's chief surgeon, Dr Alexander Macklin, who addressed the problems that allegedly accompany inbreeding. During the visit of the *Odin* in 1904 Traill Smith had been at pains to point out that 'contrary to expectation, [the islanders] are intelligent and have deteriorated neither morally nor physically by intermarriage'. Macklin's remarks on signs of deformity, however, were slightly more illuminating. 'One youth is dumb and is peculiar in manner, but … with quite average intelligence,' he noted, adding that 'one man … has stunted arms, with ill-developed hands and absence of some fingers'. The search for congenital abnormality amongst the islanders held a prurient fascination for outsiders that even today won't quite go away.

Under Rev. Rogers' direction a church was built by communal labour and opened in July 1923, despite a lack of tools and adequate materials. As well as running the school, the couple established a Boy Scout troop and introduced the island's children to football and cricket. Against this background, Rose gave birth in September 1922 to a son. To assist her she had one of the island's oldest woman, Martha Green, who had been its principal midwife for over fifty years. At the insistence of the islanders, the boy was named Edward in honour of the then Prince of Wales.

But it turned out to be a tough assignment and more than once during the Rogers' stay starvation haunted the whole community. Indeed, Rose recorded in her excellent book *The Lonely Island* that only two ships called during the 33 months they were there, the one in March 1923 being HMS *Dublin.* In the words of her husband as reported in the South African press, 'in a few weeks we should have been living like shipwrecked sailors' if the cruiser had not arrived. Eventually in February 1925 a chance arose for them to leave, and the couple grasped it eagerly. But there is a postscript. Hubert Wilkins, the naturalist with the *Quest*, had failed in his efforts to collect a specimen of the fabled Inaccessible Island Rail, the world's smallest flightless bird, but he left collecting material with Rogers, who was later able to send two specimen skins to the Natural History Museum in London for study. The species was later named *Atlantisia rogersi* as a tribute to him.

Well-heeled visitors amidst the poverty

Busy scene on Big Beach as goods are brought ashore to land by longboat from SS *Duchess of Athol* moored offshore in February 1929. (Tristan Photo Portfolio)

Meanwhile, Douglas Gane's relentless lobbying was having an effect, and by 1926 a turning point had been reached with the Colonial Office agreeing to inaugurate an annual despatch of stores, an arrangement that continued right up to the outbreak of the Second World War in 1939. On only a couple of occasions were these carried out by warships, though; more frequently use was made of large cruise-liners. The Government in effect privatised its commitment to the people of Tristan. None the less Gane's pride can be imagined when he wrote in January 1927, 'Truly a change is coming over the scene, and the hope arises that the day is not far distant when … the need for emergency relief [to the island] will disappear.'

Such relief was still vital since the islanders continued to live in pitiful conditions, plagued by rats and with hunger forever hovering in the background. Their appearance, too, emphasised the dire straits in which they survived, with the women in old-fashioned skirts like Amish folk in Pennsylvania and the men wearing such diverse hand-me-downs as naval officers' jackets or sailors' jumpers and dungarees. More than once affluent cruise-passengers rummaged through their wardrobes and donated armfuls of fresh clothing to the community. At least now there were greater opportunities for bartering. But oh! the continued isolation, epitomised by the saga of the Cunarder *Carinthia*'s visit in April 1933 when the seas were so mountainous that the ship struggled in vain to accomplish a landing. For sixteen hours the captain waited before regretfully steaming away. The islanders watched 'until all eyes were tired of looking,' as one Tristanian later wrote.

Throughout this period, Gane seized every chance to promote Tristan's exceptional character: arranging a modest display highlighting the island's unique identity at the British Empire Exhibition in Wembley, north London in 1924/25; publishing a well-received book in 1932 passionately defending his support for the rights and dignity of the Tristan community; and urging the British Museum to accept a collection, then held in America, of the 'fundamental records' relating to the establishment of the tiny settlement in 1817. Misjudged suggestions that the entire population should be evacuated were countered with Gane arguing that these annual consignments of supplies were remuneration for the services rendered by the islanders over many decades in providing succour to survivors of shipwrecks. 'Tristan da Cunha remains an ocean refuge,' he wrote, 'and the inhabitants are its keepers and should be treated accordingly.' In like spirit the Bishop of St Helena declared in 1932 that coercion was indefensible; a strong, sympathetic missionary and regular calls from ships would suffice. But the charge cannot be denied: Gane's efforts unintentionally confirmed the impression that Tristan was a charity-case living out a flimsy existence utterly reliant on the kindness of friends.

The inter-war years bring a trio of missionaries and gifts from royalty

Two key influences sustained the islanders' spirits throughout their ordeals: their deep-seated Christian belief and an unstinting affection for the British royal family. Maintaining its commitment, the SPG sent out three ministers in succession over this period – but what a curious threesome they turned out to be. In January 1927 the first of them to be dispatched was Rev. Robert Pooley, but the experience did not go well for him, and in December 1928 the SPG revealed that he was ill and appealed for a volunteer to take his place. Probably he was suffering from depression; as late as May 1931 the *Manchester Guardian* quoted him recalling his sense of 'loneliness and want of intelligent companionship.... It is positive cruelty to ignore this spot for a whole year.'

Sent to take his place in February 1929 Rev. Augustus Partridge was an ambitious individual with an impressive record of foreign missionary work. Unlike his predecessor he brought some support with him in the form of a young lay assistant, the South African-born Philip Lindsay. In late 1932 Partridge exercised his government-granted authority to set up an Island Council with Andrea Repetto's widow, Frances, in the position of Head Woman and her son Willie Repetto as Chief Man. In their separate ways the guidance from each of them would have an enduring effect on the community. It is none the less astonishing to discover that, as the law-enforcer, Partridge employed a form of punishment straight out of the mediaeval era. In 1933 one of the women on the island, Selena Green, had been accused of neglecting her home and her children, and also of abusing Frances Repetto when she attempted to clean Selena's house. Partridge had the culprit put in a set of stocks, which he himself made out of soapboxes, and even went so far as to temporarily excommunicate his victim. 'Tristan da Cunha evidently holds that cleanliness is next to godliness,' an Australian newspaper remarked shortly afterwards.

Anglican Church worship continued with full attendances at services, but for the sole Catholic family on the island the practice of their own faith was a far less pleasant experience. Speaking of Partridge the *Tablet*, a British Catholic weekly, scornfully described his style in April 1933 as 'wrong-headed and overbearing' and went on to expose his unsuccessful attempt to lobby the Colonial Secretary to refuse permission for Catholic priests to land on Tristan. '[Partridge] tried to hurt me in every way,' wrote Agnes Rogers to friends in Britain in January 1934, and so with his departure she enjoyed what she described as 'a year's peace', though she looked apprehensively towards the appearance of his successor. In the event, there was a Catholic priest on board the *Empress of Australia* in March 1935, and to Agnes's delight he came ashore and said Mass in her own home.

That successor was Rev. Harold Wilde who arrived with woeful consequences in February 1934. He was the third bachelor in a row to hold the post and over the next six years, without the potential check of any spouse, his influence on the island, although significant, was not always a benign one. His officious style soon came to the fore when he set up a storehouse under his sole control, from where he handed out rations from the island's precious stocks on a strict once-a-fortnight basis, an arrangement that caused widespread irritation within the community. 'I rule the island,' Wilde bumptiously told an American reporter in November 1937. 'The children are educated in the fear of God and directed in the paths of righteousness. I husband the supplies.' He may have felt his self-assuredness was justified but creating a power base by taking advantage of the lowish self-esteem in which the islanders held themselves hardly placed him on the moral high-ground.

Rev. Harold Wilde with the islanders' Coronation gifts ready for despatch to Buckingham Palace in March 1937. (*Illustrated London News*)

Tristan's warm regard for royalty was well founded with Prince Alfred's visit in 1867, but it was enhanced when, in response to Gane's letters to the press in January 1923 about the imminent departure of HMS *Dublin*, no less a figure than Queen Mary made a personal gift of £5 (possibly the equivalent of £400 today) for the purchase of flour. This was indeed a coup. Thereafter gifts from Buckingham Palace followed on an annual basis. In 1929 Queen Mary sent the islanders a harmonium, a gesture that would go on to acquire the status of an icon even if over its first few years it proved itself to be a bit of a dud. No-one could play the thing until Wilde's arrival almost literally gave life to the instrument. In 1935 the royal gift consisted of 6cwt (300kg) of window glass from the King, while Queen Mary made a present of a gramophone and fifty records, so it is hardly surprising that in May that year the islanders organized a 'family dinner' to celebrate George V's silver jubilee. 'I feel sure that everyone on the island will never forget their beloved King and Queen who have been most kind and devoted to their faraway subjects,' Frances Repetto wrote a short while later. Douglas Gane died in March 1935 just three months after receiving the MBE. Tristan's towering advocate, it is not over-fanciful to claim that without his almost obsessive attention, Tristan da Cunha could well have found itself relegated to little more than a footnote in the history books.

A surgeon examines an islander's teeth during the visit of HMS *Carlisle* in January 1932. (Photographer not identified)

And still the scientists kept coming, frequently expressing surprise that they found most of their hosts remarkably fit and well. Proof of this is the visit in January 1932 of the Royal Navy cruiser *Carlisle*, following which the medical specialists on board aroused world-wide curiosity by declaring that the islanders' teeth 'could only be described as perfect'. In contrast though, the perilous isolation of the community rendered it vulnerable to catching epidemic boat-colds from any passing ship. Such vulnerability became even more apparent when Donald Glass, who had left Tristan to make a new life for himself in 1932, died in a London hospital in March 1937 at the age of only 28, apparently lacking adequate powers of resistance to infection. A salutary warning, the press noted, against any policy of mass evacuation.

December 1937 saw the arrival of a Norwegian scientific expedition under the leadership of Dr Erling Christophersen, and during its four-month stay its members studied the rocks, plants, birds and marine life of the islands, as well as the health and way of life of this remarkable community of just 187 inhabitants. Key amongst the team was the Norwegian sociologist Peter Munch, whose deep interest as an observer of the community would later inspire him to spend time amongst its members during their time in Britain in the 1960s. But perhaps the expedition's most enduring legacy was the effect it had on a young engineer and trainee-surveyor, Allan Crawford. Guided by islanders, he surveyed Tristan and drew its first land map on which he recorded numerous place names that had been identified to him as his work progressed. His visit thrilled him so greatly that, like Gane before him, he developed a life-long involvement with the island community and became a leading light in promoting its interests.

Just over two years later, on 24th August 1940, Rev. Wilde left for home on the armed former-liner *Queen of Bermuda*. By then Britain had been at war for almost a year and Tristan was poised to face changes beyond its remotest imagination.

World War Two leads to a benign invasion

On the afternoon of Easter Sunday 1942 the islanders stared at the alarming sight of the armed merchant cruiser *Dunnottar Castle* approaching their shore bringing with it 35 soldiers and a significant quantity of stores. Its mission, codenamed 'Job 9', was to establish a top-secret weather-monitoring and radio station on Tristan. In that last respect its key function was to listen for radio messages from German U boats whenever they surfaced to radio back to base. So sensitive was the project that it was not until July 1945 that its details were revealed to the world, and hailed to the same extent as other classified initiatives that had helped achieve victory: ventures such as the Cockleshell Heroes raid on Bordeaux harbour in 1942, and Pluto, the D-Day fuel pipeline under the English Channel. Tristan's location had enabled a huge swathe of ocean to be monitored, thereby capitalising on its unique strategic importance for the first time since it had been recognised many decades earlier in 1816.

On Tristan, meanwhile, the islanders set about coming to terms with the military sharing their windswept plateau, building sleeping quarters, bathrooms, mess and recreation rooms, and a hospital. In due course, piped water and electricity provided by an oil-fired generator radically improved the settlement's infrastructure. To achieve such undertakings the authorities turned to the islanders for labour, paying them cash wages. In return the civilian population was permitted to buy Government stores in the station's canteen and even open savings accounts, both unprecedented experiences.

Amongst the expatriates on wartime Tristan was Allan Crawford, returning to the island as a naval meteorologist after four years in South Africa. In a gallant gesture to provide a few minutes' protection in the event of invasion he was instrumental in setting up a small but plucky armed unit called the Tristan Defence Volunteers even though, in such an isolated spot, they were never to see any action. Crawford also produced the island's first newssheet, the *Tristan Times*, which, in its early days before money was in wide circulation, cost 'three cigarettes or four big potatoes'.

Tristan Defence Volunteers on parade on VE Day in May 1945. (Janet Toseland)

In late-1943 the Admiralty resolved to commission the island as a ship, for which the name HMS *Atlantic Isle* was selected. For the christening ceremony on 15th January 1944 a West African surf boat was chosen, the whole ship's company were mustered alongside the Defence Volunteers, and Vivien Woolley, the wife of the station's commanding officer, performed the task using, in the absence of champagne, a bottle containing a mixture of rum and fruit salts. It was only in May 1946 that the Navy departed, though they left behind a welcome reserve of stores and equipment, and a permanently established meteorological station staffed by South African civilians. Now at last Tristan was assured of warships calling more frequently, each one of them bringing much appreciated supplies and mail.

There came a prince - Tristan in the 1950's

Amongst those stationed on Tristan during the war was a naval chaplain, Rev. Cyril Lawrence. Whilst there he recognised that the island-group had a potential for providing a means of self-reliance for its inhabitants through the responsible exploitation of their marine-food resources. Following his heavy lobbying of various South African fishing companies and the Colonial Office in London, he organised a scientific expedition with himself as leader to consider the practicality of setting up a fishery industry on Tristan.

On 6th February 1948 Lawrence arrived off Tristan in the tiny, decommissioned minesweeper MFV *Pequena*, and after thirty days of detailed investigations it was clear that his hunch had been correct. Amongst the dense zone of giant kelp surrounding the islands' coastlines the marine biologists discovered vast stocks of Tristan lobster amazingly easy to catch and in sufficiently abundant numbers to be a sustainable crop. However, the venture was not without sadness; the expedition team had unwittingly brought with them a severe strain of influenza which resulted in the deaths of a number of islanders, amongst them the greatly respected Frances Repetto. But back in South Africa, the commercial backers recognised that the project was a viable proposition and strategic preparations began in earnest. In 1949 the Tristan da Cunha Development Co. was established, and a lobster processing factory was built, initially for canning the delicacy and later converted to freezing in 1958. Commercial fishing was carried out by shore-based dinghies and on the outer islands by the *Pequena*, the vessel being replaced in 1954 by another small wooden minesweeper, the *Frances Repetto*. In 1952 the somewhat larger, steel-built *Tristania* was acquired, a ship that crucially was capable of providing a regular Cape-Tristan passenger and cargo service for the first time. A new era was poised to dawn, bringing economic stability to the community and enabling its members to take a more dignified place in the world.

Not that everyone saw them through rose-coloured glasses. In November 1949 the leading British photo-weekly *Picture Post* published an article which claimed that the experts on the *Pequena* had definitely not found the utopia envisaged by William Glass but rather 'a tiny society harshly divided into the few rich, a moderate middle class, and a numerous poor'. But an altogether more generous and affectionate picture around this period comes from a witty account in the Colonial Service's journal *Corona* by Lawrence's successor as chaplain, Rev. David Luard. He admired the way the women liked making their clothes from material 'of the brightest hues available, without any nonsense about clashing', naughtily remarking that some of them had even tried modern styles of dress 'which suit their bulky figures abominably'. And he was enthralled by the traditional pastimes like the 'pillow dance', noting that men not only regularly kissed women but that 'men also kiss men', adding in reassurance for his 1950s British readership that 'kissing has not the same emotional meaning that it has with us'.

By then it was clear that the unprecedented prosperity set to come with the establishment of a commercial fishing industry was bound to create unsettling challenges for the island, and the view from London was that it would be expedient to send a British Administrator to the territory. The choice fell on Hugh Elliott, a keen ornithologist and Colonial Service officer with many years' experience. Energetic and knowledgeable he quickly went about leading Tristan's local council to responsible government and, as a by-product, introducing a formalised postal service which would inevitably require its own postage stamps. This was an obvious yet inspired decision, born undoubtedly out of the earlier enthusiasm of Allan Crawford who actually prepared a set of essays priced in 'potatoes'. With the introduction of stamps to the island, philately would go on to form a lucrative source of income for Tristan for many decades to come.

In September 1955 the islanders welcomed the eight members of the Gough Island Scientific Survey (GISS), most of them from Cambridge University, who planned to make the first survey of Gough Island some 350km south of Tristan, and to study its rocks, plants, vegetation, invertebrate life, birds and seals. They came in the frigate HMS *Magpie* (formerly commanded by the Duke of Edinburgh, who was one of the expedition's supporters), together with the then Archbishop of Cape Town, two naval dentists and two expatriate families returning to Tristan. The dentists found a great deal to do on the island, for by then the islanders no longer boasted 'perfect' teeth; their way of life had changed enormously and a heightened sugar intake had led to increased dental decay. The GISS team spent six weeks on Tristan because *Tristania*, which was to take them to Gough, was delayed in Cape Town with engine trouble. In the event, this proved to be a blessing, for it allowing the party to travel all over the island, record its vegetation, and get to know the people. When they left for Gough they were accompanied by two islanders, Arthur Rogers and John the Baptist Lavarello as handymen and boatmen, later replaced by Harold Green and Ernest Repetto when the original organizer, leader and surveyor, John Heaney, was able to join the expedition in February 1956. By the time it departed in May the GISS team had completed a range of scientific studies, made the first accurate map of the island, and established a meteorological station later taken over and run permanently by the South African Weather Bureau. For the two zoologists, Michael Swales and Martin Holdgate, and the botanist (and sole Oxford University representative) Nigel Wace, this was the start of long-standing connections with the Tristan community.

Painting by Edward Seago entitled The Duke of Edinburgh departing Tristan da Cunha, showing the scene on Big Beach with islanders bidding the Duke a fond farewell. (Buckingham Palace)

In the following year, after months of planning, the day finally arrived when the Duke of Edinburgh, husband of their sovereign, paid a visit to the people of Tristan. His call was part of a four-month round-the-world tour on the HMY *Britannia* primarily to enable him to open the Olympic Games in Melbourne. Cast in the romantic spirit of the decade his string of visits to a number of exceedingly remote Commonwealth communities was seen as 'a special mission' and that he was, in the breathless words of *The Times*, highly qualified 'to bring them a message of affection from the Queen'. His support for the GISS (who had reported to him on their return to the UK) doubtless prompted a very brief visit to Gough Island prior to a day at Tristan. On 17th January 1957, and in the village that carried his own title as its name, he watched a demonstration of women carding and spinning wool and laid the foundation stone to a newly planned community centre, later to be called the Prince Philip Hall in his honour. At that stage it was little more than a skeleton of metal struts. He even participated in a brief version of the pillow dance before, with the rain falling steadily, frantic blasts from the *Britannia*'s siren lured him back on board. It had been a truly memorable day, and yet who amongst those present could have foretold what convulsions awaited their island within the next five years?

Part 3:
Volcanic Eruption and Flight 1961

Tristan's Booming Economy

Tristan's lobster industry had transformed the economy and prosperity of the community by giving record levels of revenue to the Tristan Government through royalties on Tristan Development Company profits, and a secure income through wages to the islanders for the first time in their history. By April 1961, the Big Beach fishing factory had been fully converted from canning to freezing, a process overseen by its general manager Don Binedell, who had a good working relationship with boat captains, island employees and Administrator Peter Day. Compared to work by traditional longboats, a barge with the capacity to carry weights of 3 tonnes speeded up by 50 per cent the offloading of ships, and a 35 hp tractor and 3-tonne trailer was acquired to transport loads from Big Beach to the Settlement. Islanders worked on the fishing vessel MFV *Tristania*, as well as using dinghies from Big Beach. Freezing allowed more flexible fishing, which meant that when only a few men were available, as for example during guano trips to Nightingale Island or when harvesting potatoes, five or more dinghies could still fish and have their catches dealt with by the more efficient freezing process. Men also boosted their income by government-paid building work, and this included the construction of a new house for the resident Administrator, appropriately known to everyone as the Residency. There was a radio in every home as well as a paraffin stove for cooking.

At the start of 1961 islanders were overall earning 30 per cent more compared to the year before. The island's economy was additionally boosted by increasing revenue from the sale of postage stamps, which totalled £3611 in 1960 compared to just £979 in 1959. Nowhere was the increased affluence more clearly seen than in the Island Store, still referred to as the canteen on account of its having been a former naval-station initiative during the Second World War. It was now making a profit, partly through a surcharge on alcohol sales as there was concern over the high consumption of liquor in the community. As early as 26th March 1961 turnover was running at £3203, and therefore already well ahead of the figure of £2525 for the whole of 1959.

The Prince Philip Hall had been improved by the installation of electric lighting and a new cinema projection box, which meant that a weekly film programme could be shown during the winter months. The bar made a profit, and the hall was now under the control of an effective committee of islanders in a new initiative separate from the Island Council. In contrast, Administrator Peter Day became frustrated because the members of the Island Council were reluctant to offer their opinions on any proposals for change since they felt they did not represent others, and also, since there were so few candidates, no election was ever needed. He therefore called a meeting of every one of the island men whenever key issues had to be decided, such as taxation on spirits.

Despite all this rapid progress there was a definite concern that the island was now overpopulated. This was clearly demonstrated in agriculture, since cattle numbers were rising, the pastures were over-grazed, and every winter there were stock deaths owing to starvation. New fences had helped rotate pasture more efficiently and a newly acquired Hereford bull was starting to raise breed quality. The Island Council also brought in a by-law limiting donkeys, used as draft animals to carry loads like firewood, to one per family, which resulted in several animals being culled. To improve their potato crops several of the men had started to buy imported fertiliser, recently totalling 3.5 tonnes. Some worked out that the extra money earned from fishing would offset costs, and that they would skip the Nightingale guano trips, since the fertiliser they had purchased was much more effective than the low-grade rockhopper droppings they had previously brought back.

Several islanders entered a three-year emigration trial on the Falkland Islands, for which free transport was provided. From 1959 Wilson Glass worked on a sheep farm near Port Louis on East Falkland, accompanied by his wife Maria, daughter Anne and son James who was born there in January 1961. Joseph Glass and Basil Lavarello each worked for a time on board the British Antarctic Survey RRS *Shackleton* which called at Tristan *en route* to the Falklands and aboard the Falklands Islands Co. ship *Darwin*. They had both returned to Tristan by 1961, but two more Tristan bachelors, Anderson Green and Sogneas Swain, travelled out on the *Shackleton* in October 1960 and were working on the remote Pebble Island, off the north coast of West Falkland.

April-July 1961: Tristan feels the wind of change

Peter Day completed his two-year term as Tristan da Cunha Administrator in April 1961. Following his advice, the UK Colonial Office drafted a confidential memo to act as a briefing document for new Administrators. It clarified the role of the Administrator as chair of the Island Council and explained that the islanders had a strong egalitarian tradition and did not readily accept any overlordship by their own people. Although they decided amongst themselves to give the status of Chief to one of their number, the holders of the title had in fact limited authority in the community. Indeed, there had been no elections for the Island Council, and administrators had authority to take advice from the Council or not, as they so wished. Effectively the Administrator was in charge and made all the decisions. The document concluded with the following paragraph:

> It is important that the newcomer should realise that, in spite of long-standing receipt of charity, the Tristan islanders are a proud and independent-minded people not prone, readily, to yield authority over them and resentful of outside interference in their way of life. If he realises this and treats people accordingly, he will find his appointment a unique and interesting experience.

The new Administrator ('H'admin' in local parlance) Peter Wheeler arrived on 3rd April on board RRS *Shackleton* but unfortunately, he was suffering from jaundice caused by Hepatitis B, and so he was at first confined to home. In any case the island was focussed on other health matters, as the following day HMS *Protector* arrived bringing Dr Winter of the South African Poliomyelitis Foundation who immunised all 262 islanders during his stay. Reporting to the UK's Colonial Office he criticised the absence of a resident doctor over the previous six months, especially since every islander present on Tristan at the time had been affected by a measles outbreak in September-October 1960. Islanders are still apprehensive about ship-borne illnesses which are frequently introduced.

He was, however, impressed that Tristan had a recently installed brand-new water system, which provided piped water to every dwelling, flush toilets in outside privies, and a septic-tank system which brought Tristan's houses up to an acceptable sanitary standard. Wooden privies were situated in front of all island homes and he was amused that some islanders had improved theirs by building a flat-roofed stone structure around them, which housed a kitchen sink and to which they had given the name of 'the pantry'. Dr Winter noticed in the dispensary the preponderance of inhalers for those many islanders suffering from asthma (pronounced 'H'asmere' locally). He surmised that an interplay between genetic, climatic and infective factors were the cause and thought the custom of placing rear living quarters below ground level (owing to the slope of the land) and the dampness that resulted out of that arrangement were contributory factors.

Aboard the *Shackleton,* was also a Cape Town dentist and his wife who acted as a dental nurse. Like the naval dentists in 1955 they found a great deal to do, following the coming of cash wages and sweets so easily available in the canteen. Such tempting delights were thought to have led to the poor state of children's teeth. Every one of the youngsters needed dental treatment, so even though the dental team had to work long hours it failed to complete its task. But the dentist considered the islanders to be the bravest patients he had ever attended, finding them naturally heroic and tough. Only one child whimpered a little, the rest sitting in the chair 'as quiet as mice'. The island's first dentures were later made from the impressions taken following what Dr Winter described as mass extractions, and these were brought out on the next fishing boat. Concluding his report to the Colonial Office he advised further research into the effect on health of both the increase in consumption of over-refined carbohydrate and of low-cost alcohol.

Peter Wheeler soon overcame his illness, helped no doubt by being a fit 32-year-old who had been rugby captain at Cambridge University. His colonial service had begun in 1954 when he was a district officer in Kenya, and he was joined on Tristan by his wife and their three young children. He was an agricultural specialist and took a keen interest in island farming, meeting most days with Agriculture Officer Dennis Simpson. He held regular but informal meetings with the South African factory manager, but he had a relaxed management style and, for example, generally left the teacher Ethel Bennett to tackle all education-related issues without any need to consult him.

In July 1961 Dr Norman Samuels arrived with his wife and children. Samuels was impressed that Wheeler kept his window propped open with a cricket bat, drawing the conclusion that this was a standard British colonial air-conditioning method. The island's medical and administration accommodation were brick-built but the remainder of the former naval establishment or 'station' was described by Wheeler as a random collection of wooden huts linked by covered walkways. He recalled that Island Council meetings were passive events, dominated by his briefings, and he recalled that a suggestion that whisky should rise in price was accepted without comment.

1961 was a crucial year in South Africa. The Union of South Africa had been a member of the British Empire (and later the Commonwealth of Nations) since 1910, but the National Party Government led by Hendrik Verwoerd decided to declare the country a republic following a whites-only referendum. On 14th February that year a new independent Rand currency was introduced at the rate of 2 Rand to £1 sterling. On 15th March South Africa declared its intention to withdraw from the British Commonwealth rather than be expelled for its apartheid policies. As the Cape Town-based fishing company's finances operated in South African currency, and most purchases for the Island Store were sourced in the Cape, it was decided to align Tristan's currency to that of South Africa for ease of administration. Therefore, one of the new Administrator's first duties was to oversee this currency conversion, which naturally entailed introducing islanders to the metric system of 100 cents to the Rand as opposed to the then-existing pounds, shillings, and pence of British sterling. Wheeler led a briefing session in the Prince Philip Hall and remembered the islanders taking to the new system 'like ducks to water'.

Soon islanders were readily spending money with the new currency in the island canteen, which had transferred from the factory's management to the Administrator as a co-operative venture in 1956 and was run by a local committee with the Administrator as chairman. The fledgling Tristan Post Office was supervised directly from the Administrator's Office and the old sterling stamps were destroyed in early May, after what proved to be the one and only set of thirteen Rand-value stamps was issued on 15th April. With South Africa's departure from the Commonwealth on 31st May and the authority of the apartheid regime strengthened as a direct result, this currency shift could well have become problematic, but such a fear proved to be academic as seismic events soon intervened.

On 8th August 1961, Sir Irving Gane wrote to *The Times* newspaper in London announcing the winding up of the Tristan da Cunha Fund, a source of support established by his father Douglas Gane in 1921, since the island was now no longer in need of charitable support. In his letter he explained that the island's fishing industry was flourishing, £14,000 had been paid in wages to islanders during 1960, and so its objectives had now ceased to be relevant. In recent years, the fund had provided a piano, clock, lamps and furniture for the Prince Philip Hall, pews and a new altar for St Mary's Church, and the fare for an island girl to travel to the UK. The balance of £223 was transferred to the SPG which maintained the clergyman Rev. Jack Jewell on the island. Ironically, and unknown to Sir Irving at the time, of course, the beginning of a volcanic eruption was heralded by a first tremor just two days before the appearance of his announcement, and within a matter of only a few months the fund would be hastily re-established.

The volcano awakes

The Tristan volcano, dormant for so long, resumed its activity on the evening of Sunday 6th August 1961 when most families were in their homes. The ground trembled, crockery fell from shelves, and windows rattled. A rumble was heard and then faded away. After a pause, families pondered what was going on and people gathered outside to discuss what had happened. All fell quiet until new tremors occurred on the following Tuesday and Wednesday. On Thursday 10th August six tremors were felt in quick succession accompanied by a winter thunderstorm with hail and a strong wind. The tremors were reported to Cape Town and to London. From the outset experts were hampered by their ignorance of the fact that Tristan overlies an active volcanic hot-spot and is separated tectonically from the Mid-Atlantic Ridge. An unconnected earthquake had been recorded by the Hermanus Observatory seismograph in Western Cape Province on 8th August but this was almost certainly a Mid-Atlantic Ridge tremor and as this submarine fracture is some 400km west of Tristan they are not usually felt on the island. The assumption that the quakes felt on Tristan originated from the usually benign subterranean plate movements in that more distant location, resulted in reassuring but erroneous messages from experts in London. In fact, the focus of the Tristan earthquake was so shallow that the 'long waves' (felt at the surface) were affecting only a local area and consequently not picked up by instruments in South Africa or elsewhere. The Royal Society's committee of expert volcanologists concluded that the shocks resulted from a slight settling of the ground along a possible fault line and thus assured Peter Wheeler and the community began to accept the subterranean jolts as part of life on the island.

Tremors continued and became a daily event by the end of August, with 24 being felt on one day alone. On 14th September Peter Wheeler sent a telegram to London reporting numerous minor earth tremors and subterranean 'thumps' but no corresponding tremors on Gough Island. Referring to the report from the observatory at the Cape which mentioned frequent Mid-Atlantic Ridge earthquakes, he clarified that the local shocks were the first experienced on the island in living memory. In a prophetic penultimate sentence Wheeler wrote, 'Some anxiety felt that this possibly means the re-awakening of the volcano.' He closed by asking for an expert view. It is worth noting here that his immediate contacts were Garth Pettitt representing the Secretary of State for the Colonies and Dr Shaw from the Directorate of Overseas Geological Surveys.

On Saturday 16th September MV *Tristania* arrived from Cape Town, the first vessel to appear for over five months since the visit of HMS *Protector* on 4th April.

The intervening winter had been hard, with poor shore-based fishing, so income was reduced, and families needed to draw from savings to buy groceries. Now, there were regular tremors which were increasingly regarded as a threat to the community's existence. *Tristania*'s arrival brought welcome provisions for the store, four months' mail, and a new factory manager, so there were then two company employees on the island. The vessel's captain Morris Scott, known as Scotty, was a close friend of the island, and he and his ship would go on to play a central role in the events of the following momentous month. From 16th September the number of resident expatriates had risen to a record 'station' community of 31 expats with eleven key workers and twenty family members including thirteen children. They were: Administrator Peter Wheeler, his wife and three children; Dr Samuels, his wife and two children; nurse Jean McKinley; teacher Ethel Bennett; Agricultural Officer Dennis Simpson, his wife and two sons; Rev. Jack Jewell, his wife and three children; South African factory manager Don Binedell, his wife and daughter; relieving manager Mr von Rahden, his wife and two sons; South African Meteorological Station staff Mr West and his wife; radio operator Norman Watkins; and radio technician Mr Knobel.

On Sunday 17th September, as evensong was being celebrated in St Mary's Church, the most severe earthquake so far caused a landslide on the cliffs behind the factory at Big Beach east of the village. Administrator Peter Wheeler described its effect on the church: 'Suddenly the walls heaved, the floor trembled, and for a sickening second the roof threatened to cave in.' Nevertheless, Father Jewell carried on with the service. Wheeler later wrote, 'In that moment I felt the first intimation of disaster.' Scotty came ashore and enjoyed a gin and tonic with Peter and Margaret Wheeler. At first, he scoffed at the reports of tremors, but he soon changed his mind once ashore. When later visiting island friends he recalled that it wasn't funny sitting in the islander's home since he 'thought the darn place was going to fall down', adding that some rubbish fell out from the chimney of the stone house.

Over dinner that night in the Residency, Peter Wheeler discussed with Scotty what local action should be taken to monitor the earthquakes. Scotty had read news dispatches in Cape Town about the tremors and had regarded them as something of a joke. But the intensity of that Sunday shock left Scotty surprised and somewhat concerned. In Scotty's own words, 'I was sure it was volcanic and not earth tremors under the sea, as to me the impression of explosions under the Settlement.' Wheeler added that the experience 'shook Scotty to the core'. They agreed that the task was to determine the extent of the disturbances. Accordingly, they decided to dispatch parties

Morris Scott, aka 'Scotty' in 1955 (GISS)

Administrator Peter Wheeler (Peter Wheeler)

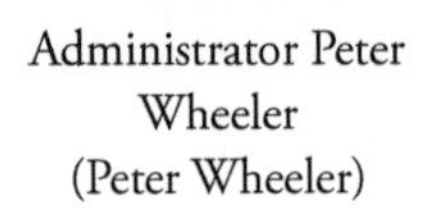

Chief Willie Repetto in 1955 (GISS)

to Nightingale and Inaccessible Islands (49 and 48 kms away from the Settlement respectively) aboard MV *Tristania* to learn if the shocks were jarring a wide area or were confined to just the main island.

The next morning, Monday 18th September, the island's Agricultural Officer Dennis Simpson gathered a small scouting party at Little Beach and from there they were transferred through a rough sea to the *Tristania* by island longboat. Scotty failed to land a party on Inaccessible as the sea conditions were so poor, so the whole group were landed on Nightingale to maintain their earthquake monitoring vigil. In his telegram to London the following day Peter Wheeler reported minor earthquakes continuing with a greater frequency and estimated that there had been five per hour the previous night, culminating in fifteen during a 30-minute period after 5am. He confirmed his plan to land a party on Nightingale to investigate whether the tremors were local to Tristan only. He was also concerned that exaggerated reports might appear in the international press and accordingly obtained the agreement of the island-based expats to cease sending unofficial reports to media links. For the first time Wheeler was contemplating more drastic action when he wrote, 'I can no longer rule out the possibility that, if they should become worse, houses may cave in and landslips may occur necessitating temporary evacuation.'

On Wednesday 20th September Peter Wheeler received a telegram from Garth Pettitt containing Dr Shaw's response. In it the latter merely confirmed that minor earthquakes were frequent on the Mid-Atlantic Ridge and, while most of the time they would not be felt, fluctuations causing shocks such as Tristan was experiencing would not be remarkable. He stated that the Tristan volcano was formed in the tertiary period of volcanic activity (65 to 1.8 million years ago) and therefore very unlikely to re-awaken. He added that there was nothing remarkable about the tremors not being felt at Gough, and finally asked for reports of any detailed movement recognised and any indication of direction. If a bottle was shaken off a table, for example, which way did it fall? Dr Shaw was then relying on a knowledge of science which had yet to fully understand plate tectonics, and by extension the identification of Tristan's deep-seated mantle plume 'hot spot'. The geology report from the Norwegian Expedition of 1937-1938 had failed to identify recent volcanic events like the Stonyhill volcanic centre which had erupted about 300 years before. This was believed to be the only Tristan geology study available in 1961 and was loaned by Garth Pettitt to Dr Shaw.

The island Administrator, faced with the reality of the threat, continued with his own common-sense research, perhaps aware that waiting to see which way a bottle dropped might be inadequate. A copy of the *Encyclopaedia Britannica* was consulted, and sensible steps were taken as a result since it was realised from the text that the shock waves ('L waves') produced from the focus of earthquakes would be felt over a smaller area the nearer to the surface they occurred. Wheeler devised a local tremor-intensity recording system which could be easily understood by all and could be used during around-the-clock watches to build up a record of the intensity and the time of each future quake. Later a Royal Society report matched these alphabetic grades with corresponding ratings on the Modified Mercalli Scale here shown in brackets:

A (3) - One that merely shook the houses slightly.
B (4) - More prolonged and louder
C (5) - One that rattled the pictures on the wall.
D (6) - Ornaments tumbled from mantlepieces; crockery rattled in cupboards

Scotty brought back the Nightingale party on 25th September with Dennis Simpson reporting that no tremors had been felt on that smaller landmass. Whilst this raised serious concern that a local eruption was now more likely, it meant that Nightingale Island could provide a safe refuge if Tristan were to become dangerous.

In a telegram to London on the 26th, Peter Wheeler reported that, during a day-and-night watch on Tristan and Nightingale from 19th-24th September, 89 'house-shaking' tremors had been felt on Tristan (as well as many minor vibrations) but none on Nightingale. He continued that 'there is now general belief that this phenomenon is local and of volcanic nature and this [conviction] will be difficult to dislodge by the long-range opinion of experts'. Wheeler was concerned about the anxiety being experienced by both islanders and expats regarding the threat that the tremors caused. As a result, he suggested that an expert 'with tact and authority' should travel out as soon as possible to make the following considered assessment: either that there was genuine danger and that the island must be evacuated, or that there was no danger and give some reason to dislodge the deeply implanted local view that a volcanic eruption was likely. Wheeler suggested that a suitable man should fly to join the Dutch liner *Tjisadane* which was due to sail to Tristan from South America in only a few days' time.

A group of islanders walked to Stony Beach to check the cattle they owned that roamed wild in the area but owing to poor weather they were marooned there for several days. During their absence there was a first instance of a 'D grade' tremor which affected houses in the village and was accompanied by an unmistakable thump followed by a prolonged shudder. When the men returned from Stony Beach they were astonished by the news since they had felt nothing at all. It was now clear that, not only were tremors confined to Tristan itself, but that they might in fact be localised around the area of the Settlement. Wheeler found this both disconcerting, since the islanders' homes might be threatened, but also encouraging, since there might be a safe haven elsewhere on the main island. Three teams were dispatched to investigate how far the tremors extended east and west. Three men climbed east via Pigbite, up the Plantation Gulch path, and made camp just below the snow line above Big Green Hill. On the way up they felt a slight tremor, but when notes were compared later it was concluded that it had been more pronounced in the Settlement. A second party went westward and climbed the mountain at Burntwood, again setting up camp just below the snow line on the Base. A third party led by Father Jewell camped near the Potato Patches.

For two days all was quiet, but canvas tents froze stiff each night on the mountain and the eastern trio abandoned their camp, though the better-equipped group above Burntwood stayed on. On the third day, a second D-grade tremor racked the Settlement. The mountain party noted it as only a slight disturbance, though Father Jewell, waking near the Potato Patches, felt it in all its intensity and, like the villagers, graded it a 'D'. It therefore became clear that it was the Settlement Plain itself which was in the centre of earthquake activity. The end of September brought another alarming development as a cliff face immediately behind the village began to break up and, with a sharp crack audible throughout the Settlement, rocks broke away from the mountainside in a landslide that crashed to the foot of the cliffs, raising clouds of dust hundreds of metres high. The mass movement continued. One rockfall killed a cow not far from the houses and the crucial water pipe running from the Big Watron spring to the fishing factory on Big Beach was cut. Consequently, an emergency pipeline had to be run from the village to keep the factory functioning. This was achieved by using a lengthy piece of hose that the *Tristania* had been carrying as part of the preparations for building a new South African government meteorological station on Gough Island. With South African permission it was now employed to supply water to the factory freezers. During this process Scotty hastily transferred 1150 cases of lobster, worth about £5000, to *Tristania*'s freezers as the factory facilities were now at risk.

On Monday 2nd October, a telegram from Garth Pettitt reported that no volcanologist with sufficient experience could travel on the *Tjisadane* and 'in any case no expert will be able to commit himself that there is no danger, as prediction of earthquakes and eruptions [are a] matter of great uncertainty'. Pettitt continued to reassure Peter Wheeler that government advisers still considered an eruption to be highly improbable, arguing that it would be more and not less likely that tremors would be felt on Nightingale if the shocks were due to volcanic activity. Commenting on the local extent of the earth tremors they were experiencing, experts thought that this could be accounted for in an adjustment of a fault line, since Tristan geology showed faulting but no active vents. Pettitt asked that checks be made on known volcanic cones to see if there was any hot ground or steam and to discover whether shocks were felt outdoors as well as inside houses. A more helpful final note added that until mid-November two frigates would be no more than three days away in travelling time, as they were on South Atlantic exercises.

A return telegram from Peter Wheeler that same day reported that local anxiety had eased and, as the south of the island was clear of tremors, this area offered a possible escape avenue if volcanic activity made the position untenable and the sea prevented evacuation. He added that no directional movement of tremors had been noted, but several dry-stone sheds had collapsed, and rockfalls near the factory had cut their water pipe. Marjorie Rogers recalled the biggest tremor so far on Tuesday 3rd October in a letter written two days later. She referred to the earthquakes as 'shakes' and noted fences around the houses falling down. (Walls are often referred to as 'fences' on Tristan.) She mentioned a warning system used by the factory whereby a man on lookout would blow a whistle to alert factory workers if a rockfall occurred, thereby enabling them to run out of the building to the relative safety of Big Beach.

Sunday 8th October 1961: the islanders abandon their homes

Although not understood at the time, by 8th October the active magma plume was already higher than all the village buildings, but within the slope behind, and about to breach the surface. With 21st-century insight previous evidence that tremors (and therefore tectonic activity) were concentrated in the village area, and that there was little lateral movement associated with tremor waves, would have precipitated an earlier evacuation, but in 1961 it seemed that it was only the Tristan community, led by Peter Wheeler, who realised the danger they were in. Early that day another earthquake (perhaps an 'E', if such a grade had been available on Wheeler's list!) caused a huge rockfall behind the Settlement. Frightened islanders saw rocks that Wheeler thought were as big as grand-pianos cascade down the cliffs, leaving an ugly scar which exposed un-weathered rock behind. The village's water supply was fed from a spring where the Big Watron emerges at the base of these cliffs. Water was (and still is) siphoned to a holding tank which acts as a reservoir, with pipes gravity-fed with the clear pure water into all island homes. On that Sunday, the concrete walls of the reservoir were smashed by rocks and the community's water supply was lost. Luckily, the boulders stopped rolling before they reached the homes below, but on inspection houses in the eastern village, nearest to the Big Watron, were found to have been damaged. Door and window frames were buckled and failed to open normally in the way they had earlier in the day. Crack lines were discovered on the walls of houses and across footpaths. Ernie Repetto lived in one of the eastern houses, close to the Big Watron, and he realised the significance of the shakes when cement cracked in the yard in front of his house.

Islanders attended evensong in St Mary's Church as usual. Later, after dark when earth movements were again felt, they took spontaneous action. Peter Repetto who ran the Island Store visited Peter Wheeler at the Residency that evening and told him that many people living in the eastern houses had decided not to remain there overnight. Wheeler went outside at about 10pm and observed a silent procession lit by torchlight, with beams bobbing hither and thither as families carrying suitcases walked slowly along, many with babies in their arms and sacks across their shoulders. He thought it was a strange and uncanny exodus, since no one had given an order to the islanders who, as if by common consent, had suddenly decided they would abandon their homes to the ominous cracks. All those living east of the Prince Philip Hall moved into western houses which so far were undamaged, making their beds on floors or in any unused corner in the homes of family and friends who, as always, made them welcome. Father Jewell carried out a tour of the houses to see if all was well and he even baptised a child whose mother was concerned about what was going on. The entire community spent a disturbed night as rockfalls continued, some accompanied by a frightening roar.

Monday 9th October 1961: no time to waste

When their owners returned to the empty eastern houses the following morning, they discovered with surprise that their doors and windows opened with ease and many cracks that had appeared in the ground on the previous day had closed. Clearly something apparently unaccountable was going on beneath the village which now had no running water to the houses. Scotty anchored the *Tristania* off the Settlement and the factory manager reported to him that islanders had not arrived for work as they had been disconcerted by the rockfalls over the weekend. They had more immediate concerns as they struggled to move back into their homes abandoned the night before. The MV *Frances Repetto* was summoned from Inaccessible Island to be on stand-by. In the early afternoon new cracks and crevices lacerated the slopes near Dockel Gulch, situated between the Big Beach factory and the most easterly house. With the *Tristania* some 500 metres offshore Scotty was asked to keep a watch from about 1pm to check the spot near the diamond navigation beacon situated there. Scotty noted that the land seemed to be rising with little cracks appearing in the grass-covered soil. He described the cracks growing higher and then falling back again, adding 'it was fascinating to watch; [the ground] seemed to be getting pushed out from the mountain side'. Some islanders themselves described one fissure developing into a crevice some 3 metres deep in which a sheep was trapped. The bottom of the cleft then pushed upwards, and the sheep escaped.

Peter Wheeler was summoned from the Administration Office to the scene by Chief Willie Repetto. Wheeler described one crack where the ground had separated and, while one side remained stationary, the other had lifted more than 3 metres, creating a vertical cliff on top of which teetered an enormous boulder. It was the sight of this mighty rock, precariously balanced and threatening to teeter over and tumble down onto the village below, that led Wheeler and Chief Willie to believe that the end had come. Clearly there must be some enormous pressure forcing up the ground. Time was running out, so Wheeler decided to evacuate the island, but this would have to be postponed since a safe embarkation by longboat that late in the afternoon to the tiny *Tristania* would be impossible. He therefore returned to the Settlement and struck the village gong, an old naval torpedo shell, to summon the island men to a meeting at the Prince Philip Hall. At about 5.30pm he announced that it would be wise to get away from the immediate danger in the village. He told the men to return to their homes, gather everybody together, get their womenfolk to pack warm clothes and blankets, and immediately start an exodus to the Potato Patches where temporary shelter would be sought for the night. There was no discussion or questioning, no need to reach an agreement; everyone just got on with a calm and orderly evacuation of their homes. Wheeler later added that it was just a time for action and the people were marvellous.

He hurriedly wrote an emergency telegram to Cape Town to alert the Royal Navy and this was sent as usual by the radio operator Norman Watkins using Morse code. There was a strict and clearly understood protocol for telegrams used in the British Colonial Service and Armed Forces, and such messages were ranked according to their importance in this order: Normal, Urgent, Most Urgent, Immediate and, finally, for extreme emergencies: Clear the Line. The latter may be used occasionally by Prime Ministers, but the Simonstown receiving station had never received a 'Clear the Line' message before, so all those involved were on full alert. Immediately the despatch was conveyed to the Royal Navy indicating that there was grave danger on the island, that the decision to evacuate had been made, and that this was a request for the urgent assistance of a warship.

A second telegram to London followed at 5.55pm:

> EMERGENCY. Ground cracking badly near Settlement. In one place ground has lifted 10 feet [3 metres]. May have to evacuate to Nightingale. Have requested Navy Cape Town to close frigates on Tristan. Please send all messages in clear: no time for decoding.

A reply confirmed that the two frigates were in dock at Simonstown and not on exercise in the South Atlantic as the previous telegram received a week earlier had indicated. Therefore, help was more than the three days away as a ship would need to be prepared to sail and would take longer to reach Tristan. Wheeler remained in close touch with the *Tristania* and made a request for the fishing vessels to stand by to evacuate all the community to Nightingale Island the following day. Scotty continued to keep watch with binoculars just offshore. He saw the ground around the original fissure get bigger and bigger, forming by the time darkness fell a kind of little hill about 15 metres high, by which time he thought the cone looked like a blow hole. Just before dark he watched the diamond navigation beacon topple over.

Meanwhile the trek westwards continued in the twilight. The Potato Patches are situated some 3km west from Hottentot Gulch, reached by a rough track which twisted up and over the Valley behind Hillpiece. The island had just one tractor which was driven by Agriculture Officer Dennis Simpson and he made a series of trips to and from the Patches. By his actions he was able to carry some of the very old and infirm villagers directly there, though they experienced an unpleasant ride as they bounced along the road huddled on the trailer behind the vehicle. Donkeys helped carry belongings. Mary Swain's family dressed in overcoats and mackintoshes to fend off the rain took their rifle along just in case. Mary remembered her husband Fred being very tired and she shone a torch to show him the puddles and ponds as they went along the way. Others suffered severe grazes and bruises as they hastily trod the stony pathway in the wet gloom. The walk was hard for some and worrying for everybody. With dusk approaching there was a cold drizzle to accompany the sad procession, as they faced an uncomfortable night and an uncertain future. Sister McKinley saw the last tractor trip arrive by torch light with Dennis Simpson's fingernails worn to the quick as he struggled to keep the vehicle on the path.

The buildings at the Potato Patches were in those days confined to three stone huts plus a few small sheds designed to store seed potatoes and the hand tools used in the walled vegetable plots. Today there are well equipped 'camping huts' that have beds, stoves that use bottled gas for fuel, and ample space. These did not exist in 1961, and so on the night of Monday 9th October it was only the old islanders, people with infirmities, and mothers with babies and small children who had a roof over their heads. It was a chilly experience, closer to freezing than in the well-insulated stone cottages back in the village. Mary Swain thought the crowd in her family hut were like 'sardines in a tin, cold, miserable, with no drink, no supper, no nothin''. The island's only two tents were erected at Below the Hill, and in these the old and sick huddled together with the Glass family. These included Liza, her son Edwin (Spike), his wife Monica, and their daughter Sheila and baby Conrad. Men crouched in the ditches, behind walls, and in empty oil drums seeking protection from the biting wind sweeping in from the sea. Young men, including Cheseldon and Basil Lavarello, and Adam Swain, stayed awake and kept up a watch around the huts. Peter Wheeler, his wife and their three children huddled in a canvas mail bag at the bottom of a ditch to keep out of the bitter wind, making sure that the portable radio Wheeler had brought remained dry and able to receive any messages from Scotty.

When the *Frances Repetto* arrived from Inaccessible, Scotty sent it east off Big Beach to stand by for the planned departure next day. He took the *Tristania* westwards and lay off the Potato Patches, far enough out to see the Settlement and the menacing new mound beyond. He could also relay messages about any new developments to Wheeler and liaise about departure next day. During the night Scotty and his crew kept a strict watch in the direction of the Settlement and the growing dome behind it, none of them sure of what was going to happen. At 9.20pm a telegram was sent to London from Simonstown:

Following reports of severe earth tremors and at request Administrator intend to sail *Leopard* at 20 knots for Tristan da Cunha. Expected time of arrival 13th October with relief supplies and to assure communications should situation deteriorate further.

A sense of foreboding permeated all ashore and on the two fishing vessels as midnight approached. Talk ceased and the islanders dozed, but no one slept peacefully. It was at least reassuring to see the lights of MV *Tristania* bobbing up and down offshore.

Tuesday 10th October 1961: the fateful day – Tristan is evacuated

At 2am in the morning of the 10th a volcanic eruption was seen from aboard the *Tristania*. Scotty was woken from a snooze at 2.15am and alerted to a fire visible in the direction of the Settlement. He set off warning flares, but these were unseen by Peter Wheeler. Someone shook Wheeler into consciousness, and he stumbled to the radio. Scotty was on the other end. 'The bubble's blown open, Peter,' he said simply. 'It's pushing up rock and hot cinders and belching smoke.' As Dr Samuels later recalled, 'I shall never forget the sight of men huddled around our small radio transmitter, talking to the *Tristania* and hearing that our volcano was now erupting in earnest.'

Peter Wheeler met with Chief Willie Repetto and it was agreed that the longboat coxswains and crews would return to Little Beach, launch their boats and bring them back to Boatharbour Bay where all those ashore would be taken on board and transferred to the fishing vessels. Boatharbour Bay has a sheltered beach and lies immediately west of Hillpiece. It can be approached safely by taking a path down the low cliffs from the Patches Plain and then along the sandy Runaway Beach. It offered the best prospect of keeping most people well away from the erupting volcano and allowing them to escape the island *en route* to Nightingale Island and safety. At daylight, the *Tristania* steamed back to the Settlement and to the erupting volcanic cone, which had now grown to about 20 metres high and nearly 50 metres in diameter. It looked higher because its central vent was situated on steeply sloping ground. Scotty described it as 'quite a sight' with smoke and red-hot rocks falling or rolling down the sides from a summit crater. In the chill half-light of dawn, Wheeler assembled the men and they walked together back along the track over the Valley and past Knockfolly Ridge, and as they climbed out of Hottentot Gulch and mounted the rise beyond it the village came into sight. Just on the far side of the houses white smoke swirled from the mouth of new volcanic cone. Here was the agent of their undoing – a raw, open, fiery wound in the earth. Weary and drained of emotion, they could only stare at it in silence. In Dr Samuels' words, there was 'a sulphurous smell in the air and red-hot rocks and flames were being spewed from the crater. The rocks tumbled down the sides of the new cone flaming as they fell, and smoke hung overhead, every now and then rising in a big puff like the mushroom cloud of an atomic bomb.'

The news of the momentous events unfolding on Tristan was first announced in the morning edition of the *Cape Times* with the headline, 'Warship off to Tristan after New Shocks'. The article began, 'After reports from the island's Administrator of more severe tremors on Tristan da Cunha, the Royal Navy frigate HMS *Leopard* will sail for the island at 8am today.' There was no news of the evacuation because, when pressed for details of the situation, an official simply replied, 'Reports we have received are sufficiently serious to warrant sending *Leopard.*' Nevertheless, the international media were now on the alert and Tristan da Cunha would soon be headlines in newspapers, radio, and TV across the world. In fact, HMS *Leopard* had already sailed earlier that morning at 4am and so this is just one of any number of facts about Tristan itself, the eruption, and various events surrounding it, which were wrongly reported across the international media. Setting out at its maximum speed of 20 knots (37kph) the ship was due to arrive on 13th October, but at that speed its fuel consumption would be very high and might entail an immediate turn-round.

Wheeler and some of the men hastily salvaged portable valuables from the houses while others passed through the eerie emptiness of the village and down to Little Beach. There, a mere 200 metres from the erupting volcano, they pulled down four longboats from the grass bank and launched them into the sea. Very soon they were rowing past the Settlement towards Boatharbour Bay where the others were waiting to be evacuated. There is a reef offshore from the bay where in certain circumstances waves break turbulently on submerged but shallow rocks which islanders call 'blinders'. At that point it was the most experienced coxswain, Lawrence Lavarello, son of Gaetano Lavarello who had introduced longboats to Tristan in the 1890s, who took responsibility for attempting the first landing. His boat was known simply as *Longboat* but was also nick-named *Guttersnake* because at some 10 metres in length it was the longest boat. Later it would be called *British Flag*. Timing a run to shore has to be exact, and on Lawrence's signal his crew rowed hard to catch a breaker which then fired them towards the shore, where they made a safe landing on the grey sand beach. Here the first boatload of refugees was picked up with older islanders identified as a priority. One of these was Lawrence's mother Jane who was the last surviving granddaughter of Corporal William Glass and the island's oldest resident. *Canton*, controlled by coxswain Thomas Glass, also made it into the bay and both boats carried their precious cargoes through the blinders, across the choppy water and out to the awaiting *Tristania*. Scotty thought getting more people aboard would be too dangerous as there was too much swell to allow a safe transfer from longboat to ship. Further longboat trips were therefore cancelled. Peter Wheeler thought *Longboat* missed a blinder by just half a metre and believed it would be madness to continue using the bay. Scotty towed the four longboats back to the Settlement and laid off Little Beach, which could well offer a safer escape route for everyone else. However, Little Beach was a mere 200 metres from an erupting volcano.

Islanders and expatriates now had to face the return journey to the village. The flight of the evening before was reversed, but this time around they all had to start from the beach at Boatharbour Bay, with a stiff walk along the sand, up the steep cliff path and then along the more gently sloping grass sward before they reached the Potato Patches. From there the road (on Tristan even footpaths, and certainly rough tracks, are called 'roads') led back to the village and beyond to Little Beach – probably a good 6km all told. It was too much for some. Octogenarian Annie Swain could walk no further than Runaway Beach, so Sidney Glass and Cyril Rogers caught a donkey, helped Annie onto its back and guided the animal with the elderly lady up top all the way to the Settlement. Although the emerging volcanic cone was threatening, there was no lava flowing from its crater. However, the rocks cascading down the cone's sides were red hot. White vapour, often described as 'smoke', emitted skywards and sometimes towards the village, and this contained poisonous sulphuric gases which could cause suffocation. Nevertheless, people returned to their homes to retrieve their valuables. Lisa Glass collected a few more clothes, but a breakaway group of women scrambled down Hottentot Gulch, and then along the boulder beach past Hottentot Point to Little Beach, avoiding the village completely and the view of the ominous eruption beyond.

View from a longboat during the evacuation of Tristan da Cunha on 10th October 1961 showing the new volcanic cone behind a rising conical mound which itself would soon be dislodged and engulfed by emerging lava. Below left is the Big Beach fish factory. (Geoffrey Dominy)

It was noticed that bizarrely the waters of Big Watron, the only reliable water supply on the Settlement Plain, had heated up and turned milky in appearance. As the island's precious water supply looked contaminated by the eruption, there were no second thoughts about the wisdom of immediate evacuation in what had become an untenable place. Dennis Green lived in the house nearest the eruption and yet he risked returning home to secure what he considered to be the most precious of all his remaining belongings. Bizarre as it may sound to those who own high-value possessions, these were three fresh oranges, a luxury that had arrived from Cape Town in September, which Dennis was determined would be eaten later by his wife Ada and two-year-old daughter Valerie. He then dashed down the banks of the now hot Big Watron to the safety of the waiting longboats, to re-join his family and await their turn to abandon their island home. Dr Samuels witnessed the refugees' return to the village, all tired, dirty, and resigned, but he did not hear one word of complaint. All accepted their fate as a matter of course. At Little Beach, the four longboats operated a shuttle from mid-morning. Most of the islanders were aboard the two ships by 11.30am and Peter Wheeler was the last to leave in the early afternoon, having made sure the evacuation was complete and all the 295 refugees were accounted for. In the case of each of the vessels this was a remarkable achievement. The *Frances Repetto* had never carried passengers at all; it was merely a converted 316-tonne wooden motor-minesweeper. The *Tristania* on the other hand was a 628-tonne steel-built fishing boat, but even then, it could normally accommodate no more than twelve passengers.

Three longboats with evacuees being ferried from Little Beach to the MV Frances Repetto during the evacuation of Tristan da Cunha on 10th October 1961. The newly emerged volcano can be seen smoking behind on rising ground between the village and Big Beach fishing factory, shown far left. (Geoffrey Dominy)

From aboard the *Tristania* people saw a disturbance at about 1pm occurring in a bog just below the Big Watron spring and about 200 metres west of the growing volcanic mound. Scotty observed a pall of yellowish-brown vapour rising over the marshland. It is likely that this strange occurrence was caused by hot magma below vaporising water at the bottom of the bog, and the effect of this disturbance was to discharge mud from the swamp onto surrounding grassy slopes and to dislodge and disgorge boulders. Almost certainly this same process further east accounted for the hot milky waters flowing in the Big Watron and over the cliff waterfall and onto Little Beach.

Reflecting on that afternoon Peter Wheeler later said that 'as the Settlement slipped below the horizon, the islanders watched without a tear. It was, perhaps, enough to be alive'. The emotion was undoubtedly one of relief that no one had been injured, and that everyone was safe, though these feelings were offset by the realisation that the village appeared to be tearing apart. The previous 24 hours had been all about safety and escape; there had been no time to rationally prioritise what luggage should be taken, no chance to secure valuables in offices and homes, and no time to consider the remaining livestock, dogs and cats. On Tristan that day there were 264 islanders and 31 expatriates, and some comfort could be drawn from the fact that a successful departure had been secured for every one of them. Wheeler was able to set up an office aboard the *Tristania*, from which he could co-ordinate the evacuation and liaise with London and Cape Town using the ship's radio. His first message to the Colonial Office in London at 4.50pm that day read, 'First phase of evacuation completed. All people off island. Carried out successfully only because of splendid co-operation of *Tristania* and *Frances Repetto*. People behaved with admirable calmness throughout operation.'

At about 5pm the ships arrived at Nightingale Island. Nightingale is the smallest of the four main Tristan da Cunha islands and frequently visited by longboats to gather birds' eggs, guano and fat for cooking, as well as capturing many thousands of birds for food. Scotty had loaded two longboats aboard the *Tristania*, and the *Frances Repetto* had towed a third. The main landing rock on Nightingale was the locality Scotty chose to approach since it offered some shelter from a westerly swell. The spot consists of a smooth wide convex rock leading to a cave in which longboats can be drawn up and secured so that they are not washed away in an easterly swell. Most islanders went ashore to spend the night on dry land, but it was simply not feasible for the elderly and infirm to safely transfer to Nightingale. The three longboats shuttled those who were fit from the two fishing vessels, which by now were anchored close offshore since a safe deep-water anchorage is here much closer to the shoreline than at Tristan. The longboats cautiously approached the landing place, bow first, allowing the most agile to jump onto the slippery rock, get a foothold and hold a rope fast, while the coxswain at the helm held the steering oar steady to ensure the fragile boats remained at a right-angle to the spot and the incoming swell. People were helped, hand-in-hand, each having to make a leap from the bow of the longboat, as it constantly rose and fell with every surge. Even on a perfect day, landing on Nightingale is not for the faint-hearted. Behind the landing point there stood a few huts crudely built on the bare earth, which was typically a peat soil riddled with the burrows of seabirds including petrels and nightbirds. Their walls were of volcanic boulders, simply piled and usually without mortar, with roofs of corrugated zinc-coated steel on a wooden frame. More accommodation than at the Potato Patches, but not enough for everyone who came ashore.

Ashore there was fresh water, which was gathered in butts caught by roof gutters since the streams on Nightingale are polluted with guano. A contingency store left by islanders was also put to use. This consisted of tea, coffee, biscuits, potatoes, and petrel fat. It was quickly requisitioned as the largest group ever to land on Nightingale settled in for the night. Soon wood smoke was seen rising from the hut chimneys as fires were lit. Petrels were starting to return to breed from their northern hemisphere feeding grounds and these birds were readily caught to fry in pans of last season's petrel fat alongside sufficient quantities of 'tatty cakes' (boiled potatoes, which are mashed and then fried). So, islanders enjoyed a first good hot meal more than a day after evacuating the village. There was one particularly difficult job for the men to do, and that was to shoot all the dogs that had been brought with them to Nightingale, and thoughts turned to the dogs remaining on Tristan that could cause havoc with the stock there unless they were dealt with. Many older and infirm islanders, together with the expatriate women and children, remained on the two fishing vessels overnight. The *Tristania*'s crew vacated their cabins to accommodate women and children especially those of expat officers, and evacuees also crammed into other spaces on board including the processing and mess rooms, as well as on the open decks.

A telegram received at 5.30pm from the Colonial Office in London and headed 'Emergency' offered little to help Peter Wheeler as he considered his options of what to do next. The telegram confirmed that the authorities were aware that an evacuation to Nightingale was being attempted adding that this 'can only be of limited duration and if in your opinion [the] position warrants it total evacuation should be made to Cape Town where arrangements will be made for continuation of evacuation to United Kingdom'. The message ended with the words, 'All here deeply concerned and anxious to give any possible help.' Peter Wheeler made crucial arrangements with amazing speed. He was in contact with the Royal Dutch Interocean Line ship *Tjisadane* which was scheduled to call at Tristan the following morning. Instead, he arranged for the vessel to change course to Nightingale to pick up the refugees and transport them to Cape Town. His telegram to the Colonial Office at 8.50 that evening informed them that the *Tjisadane* would arrive at 9.30 the following morning and that it had offered to take off the whole population to Cape Town. Wheeler requested immediate instructions from London on whether to accept this offer. He had already made up his mind to send all the expatriate women and children on the ship. He ended the telegram, 'Imperative answer received here tonight.' However, he soon regained his resolve and made the decision without waiting for permission. His next emergency telegram to London followed in ten minutes at 9pm. It simply read, 'Am evacuating total population to Cape Town on *Tjisadane* and *Leopard.* Please arrange reception accommodation Cape Town.'

On Nightingale by about that time Father Jewell, Dennis Simpson and Dr Samuels settled down to their first proper meal for 36 hours consisting of thick slices of bread, sausage, guava juice and fruit. Peter Wheeler came ashore to report that the *Tjisadane* would arrive in the morning which raised everybody's spirits. He checked that all was well and then returned to the *Tristania*. The huddled refugees both aboard the ships and on Nightingale bedded down for the second night of their evacuation. The island is never quiet. Millions of seabirds breed there and many of the smaller burrowing birds are nocturnal, often attracted to ship lights. Many a night ashore is punctuated by a dream or nightmare connected to the sudden appearance, as if by magic, of a nightbird mistaking a sleeping bag for its burrow.

Tristan islanders still speak of the miracle (they may also call it an 'Act of God') of their escape without harm from the eruption of October 1961. It is worth here considering the elements which made up what was at its most basic a series of lucky circumstances.

- The eruption itself was gradual enough to allow safe passage in good time to the Potato Patches, and the successful evacuation the following day. Often eruptions are much quicker, more violent and accompanied by emissions of ash which roast and suffocate.
- 10th October provided calm conditions at Little Beach to enable longboats to take off the population safely. On most Tristan days that is not the case.
- There were two fishing vessels already at the island. If the eruption had occurred between April and August that year, no ships would have been available locally to render assistance.
- Quite by chance, a passenger liner was due to call on 11th October to collect six passengers from Tristan for onward travel to Cape Town. Furthermore, the ship was virtually empty, carrying only twenty passengers with room for 400.
- Islanders have been heard to say that in Peter Wheeler they had an Administrator who made good decisions and acted decisively.

Wednesday 11th October 1961: Departing from the islands

Overnight on Nightingale there were informal conversations amongst islanders about their predicament. Apparently, some of the younger island men thought it would be worthwhile for a group of them to remain on Nightingale and not be evacuated aboard the *Tjisadane* or HMS *Leopard*. Later these discussions were recalled to the sociologist Peter Munch in the UK and recorded in his book *Crisis in Utopia*. Their thought was to return to Tristan da Cunha by longboat to take care of the stock by killing the sheep dogs which otherwise were likely to form wild hunting packs, and to maintain the houses. However, these discussions were entirely unofficial; there was no gathering of Island Council members, and crucially no communication whatsoever with Peter Wheeler, who remained aboard the *Tristania* overnight. When later summarising the situation, Wheeler confirmed that he had decided to evacuate every individual onto the *Tjisadane*, and to take personal control of any Tristan salvage operation using the crews from the fishing boats, who were eager to retrieve company property from the factory, and the *Leopard*'s crew, who were due to arrive 48 hours later. No islander offered to join Wheeler and no alternative plans were suggested by anyone else.

Islanders gathered at the rock landing place on Nightingale Island awaiting transfer by longboat to the Dutch liner Tjisadane on 11th October 1961.
(Tristan Photo Portfolio)

A child being gently lowered from MV *Frances Repetto* to join islanders in a waiting longboat on 11th October 1961 before being transported to the Dutch liner *Tjisadane* for their passage to Cape Town. (Geoffrey Dominy)

The Dutch liner *Tjisadane* had previously transported a full passenger complement of Japanese and Chinese migrants to Brazil and was currently undertaking a return voyage to Asia via Cape Town. The Royal Dutch Interocean Line would therefore have been pleased to secure the six extra fare-paying passengers due to be loaded at Tristan da Cunha that morning. This group comprised a family of four expatriates and two young island women scheduled to travel on to the UK for training as nurses. The switch to nearby Nightingale Island, which would result in an increase in the ship's payload to over 75 per cent of its capacity, was consequently an attractive arrangement for the company. At about 9.30am the *Tjisadane* under the command of Capt. Giel arrived off the island and soon the longboats started to transfer the evacuees over to the 9,300-tonne liner. Dinghies from the fishing boats also operated a shuttle to transport those aboard the *Tristania* and the *Frances Repetto* to the ship. The Dutch vessel let down a gangway which, though steep and unsteady, offered easier access than the swaying rope ladders alongside the fishing vessels and proved to be a relatively comfortable way of boarding. Better still – the cabins were more luxurious than the islanders' own homes. As a final task, the three longboats were lifted aboard and stored on the ship's

A longboat with a group of Tristan refugees aboard being rowed towards the liner *Tjisadane* on 11th October 1961. (Geoffrey Dominy)

Islanders gather on the deck of the Dutch liner *Tjisadane* on 11th October 1961 as the ship offers them what was assumed to be their final sight of Tristan da Cunha, the home they had abandoned only the previous day. The ship is here off Hillpiece and the Settlement comes into view, the people are anxiously looking ahead to see how the eruption has been threatening their homes. (Tristan Photo Portfolio)

decks, and the liner set sail after a loading operation that lasted about four hours. Peter Wheeler thought the crews of the two fishing vessels had done a 'marvellous job' to ensure everybody boarded the ship safely.

The *Tjisadane* steered north-east so that islanders could view the home they were leaving that afternoon for perhaps the last time. By then highly emotional they lined the decks on the starboard side as the ship passed Hillpiece and the now-deserted Settlement came into view. Little was said though many tears were shed. The sight of the new volcanic cone, blowing out sulphurous grey clouds charged with gas only confirmed the wisdom of their escape, as did the red-hot chunks of rock which they could see cascading down the slopes. Passing Big Point and Sandy Point the vessel struck eastwards towards Cape Town and an uncertain future. Even though the Dutch liner had apparently enough empty cabins for all a number of these contained packages of freight, and so some islanders were accommodated in cargo space. Dennis Green was one of the men who as a result was obliged to sleep not in a bed but in a hammock.

Peter Wheeler remained aboard the *Tristania* which followed the *Tjisadane* from Nightingale to Tristan. By doing so he was able to liaise with the Royal Navy and officials at the Colonial Office to plan the salvage operation on the island now that all its resident population and the various expatriates were safely being looked after *en route* to Cape Town. An early telegram that day had confirmed that HMS *Leopard* was carrying extensive emergency stores which included 3000 man-days of dry provisions, items of clothing, twenty tents, ground sheets and tarpaulins, and three portable radio sets. The frigate was also bringing timber, paraffin, lamps, torches and medical supplies, but all these were redundant, now that the people were safe. The *Leopard* was still proceeding that morning at around 35kph based on the expectation that the ship was conducting a rescue operation. When Peter Wheeler informed its commander that *Leopard* would arrive 'after the party was over', thereby changing the focus of its mission, the ship reduced speed to conserve fuel and extended the limit of the exercise. A decision was made by the Colonial Office and the Admiralty that HMS *Leopard*'s assignment should now centre on bringing back or destroying confidential documents, and carrying back to Cape Town the Administrator Peter Wheeler, now the UK's senior official of an island devoid of any resident inhabitants.

Communications between the *Tristania* and the frigate were not always easy because the fishing vessel was experiencing battery problems with its radio equipment, so at times contact was lost. Cdr Hicks-Beach aboard *Leopard* also 'cursed the amount of time and battery power being devoted to press messages, which cluttered the ether and precluded obtaining answers to urgent questions'. By now all the world's media wanted to know exactly what was going on in the South Atlantic. By early evening the future of the Tristan islanders was already being considered as the Colonial Office gave thought to arrangements for the reception of the islanders in the United Kingdom. A telegram to Peter Wheeler at 6.45pm went on to say:

We are assuming that it will not be feasible or desirable for islanders to return to Tristan and that United Kingdom Ambassador Pretoria will wish to arrange their onward passage to the United Kingdom as soon as possible. We are already in contact with Government Departments here both as regards reception and resettlement.

The Colonial Office also instructed Peter Wheeler to fly ahead to London after he arrived in Cape Town to advise on future planning. At that very moment, the islanders themselves were looking back towards Tristan in the evening light from the stern of the *Tjisadane* and preparing for the first night on their long journey into the unknown. Peter Wheeler's final act on this day was to send a poignant telegram to the Colonial Office which encapsulated the responsibility and danger he could see himself facing in the next day or two:

> Consider my duty here. Am aboard *Tristania* waiting for *Leopard*. If beach conditions warrant it intend going ashore tomorrow with party of volunteers to do certain jobs. There will be danger, but I think the risk worth taking [in] view of your cable via *Leopard* [which indicated reduced speed and later arrival]. If anything goes wrong trust will be exonerated of responsibility. If no reply from you will act if possible.

Thursday 12th October 1961: The hazardous salvage operation begins

Wheeler's message was received by the Colonial Office shortly before one o'clock on the morning of Thursday 12th October and its reply sent at 8.30am shows that the officials in London had not fully understood its significance and did not want any risks to be taken. Their response concluded, 'Do not attempt to land on island until arrival of *Leopard* or further instructions,' but the message was far too late for Scotty and Peter Wheeler. Engineers and fishermen from *Tristania* had already gone ashore at Big Beach and started to strip the factory and get all the fishing equipment off the island. Calm conditions for landing prevailed and the new volcanic cone was changing more slowly after its initial rapid growth, so it seemed safe to undertake these tasks. Little Beach gave direct access to a track leading into the village and close to the Administration Office which held the island safe. Effectively, this was the Tristan bank, since it contained the stock of cash as well as precious codes and ciphers which in the wrong hands could undermine Britain's intelligence activities around the world. From that office wages were paid to government employees and people could make purchases, including postage stamps. Unfortunately, in the haste of events two days earlier, the key had been lost. Undeterred, Wheeler and two Afrikaner officers from the *Tristania* got hold of a sledgehammer, rolled up their sleeves and took turns to 'smash hell out of the safe'. The crude operation was successful; the door was broken open and codes, ciphers and about £5000 in cash were secured.

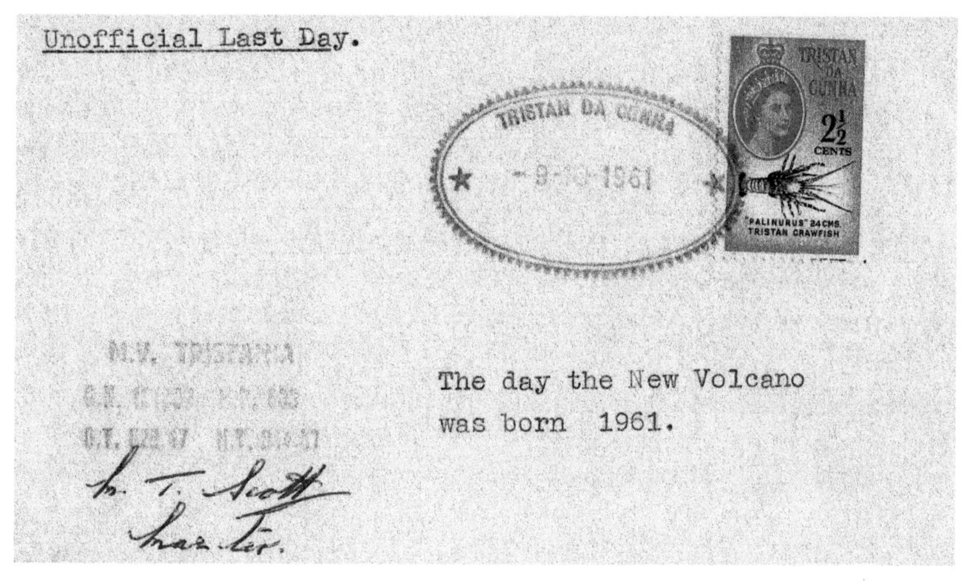

An envelope assembled by Captain Scott bearing a 1961 Tristan da Cunha 'Rand' stamp and franked by him with an acquired official hand-stamp, thereby enabling him to create a cover commemorating the start of the volcanic eruption on 9th October 1961. It was to prevent more of these unofficial covers being produced that Peter Wheeler felt impelled to destroy so dramatically the remaining stocks of the island's stamps on 12th October. (Robin Taylor)

Nevertheless, the room was now in a state of complete disarray. Since the Administration Office in those days doubled as the island's Post Office its precious stock of stamps was kept in a filing cabinet, and this was also locked with the keys missing. With three strong and healthy men available it was decided to lift the entire cabinet and load it onto the trailer of the island's tractor. Wheeler well knew of the increasing interest in Tristan's postage stamps and the strong commercial business that accompanied them. He also knew of Scotty's strong philatelic interest, and indeed regarded him as a 'manic stamp collector'. Scotty was in regular contact with several stamp dealers, and he frequently created postal items, carefully addressed, stamped and franked, to send to correspondents who would build up a collection later to be sold at many times its original cost. So, the current stock of recently issued Rand stamps with designs featuring Tristan marine life, was now likely to be of exceptional value to collectors, especially if it fell into unscrupulous hands. Wheeler, therefore, decided with deep regret that the cabinet containing the entire reserve of mint stamps should not go aboard the *Tristania*, but would have to be destroyed. The tractor and trailer proceeded down the track to Little Beach, and the filing cabinet was carefully positioned across the bow of one of the dinghies, thereby helping to balance the weight of the three men behind. With the officers rowing, Wheeler, by now halfway back to the *Tristania* and in deep enough water, tipped the container over the side into the ocean. He waited, but it seemed watertight, and it refused to sink. Luckily, he had a revolver with him, and he proceeded to shoot holes in its side to ensure that it was consigned to the deep. His action proved effective, and with the sound of loud gurgling this symbolic act by the island's Administrator consigned its entire contents to Davy Jones's Locker, ironically creating in an instant an immediate profit for those collectors who already held the few mint stamps and covers of the new series already sold.

The British Embassy in Pretoria made onward travel arrangements for the Tristan islanders and confirmed just after 3pm that day that the Union Castle Co. had offered to accommodate all aboard the liner *Stirling Castle* which was scheduled to leave Cape Town for the United Kingdom on 20th October. The ship would be available at the dockside, so accommodation in Cape Town would not be needed since the islanders could transfer immediately from the *Tjisadane* on arrival. Other confirmed arrangements included the local Red Cross preparing to give any help required, and the British Embassy officer in Pretoria, Peter Lewis, was designated to travel to Cape Town to co-ordinate the plans for reception. At no stage was South Africa considered as an alternative refuge for the Tristan islanders. Their status in the rapidly strengthening apartheid regime was an uncertain one, unlike that of their ancestors who had been able to settle without any difficulty in the expanding Cape Colony as fully acknowledged British subjects. The Permanent Under-Secretary at the Colonial Office, Sir Hilton Poynton, confirmed that the islanders were exceedingly loyal subjects of the Queen, that they had few links (other than commercial) with South Africa, and that 'they would fall the wrong side of the South African colour bar though their dominant strain is European'.

At 5pm Peter Wheeler informed the Colonial Office in London of the successful landing and the recovery of cash and documents. He reported the hazardous company salvage operation close under the erupting cone in which *Tristania*'s crew had recovered much of their beach equipment. He confirmed that the Settlement was further away from the eruption than the factory site and would only be threatened in the next few days if there were to be a sudden and violent increase in activity. He then checked whether HMS *Leopard*, which was expected the next day, could be used to recover as much as possible from the village, or whether it should return to Cape Town, since its mission to evacuate the population and to secure official documents had already been achieved. In an artful finale to this message, designed to push the authorities into committing to a full salvage operation, he ended with the words, 'If you instruct me to return immediately, am I to get navy to demolish settlement by gunfire? Require answer before tomorrow.' That very night he was to receive a positive answer via the fast-approaching warship. HMS *Leopard* passed the *Tjisadane* during the evening and Cdr Hicks-Beach contacted the *Tristania* by radio to clarify that the warship would indeed remain at Tristan if conditions permitted in order to help salvage the islanders' personal belongings.

The Tristan volcanic eruption, the successful evacuation of the islanders and speculation about their future was by now a major feature on television and radio news bulletins and in newspapers across the world. From Balmoral Castle Queen Elizabeth II sent a message headed 'Tristan da Cunha Disaster' which read,

I and my husband send our deepest sympathy to the inhabitants of Tristan da Cunha in the tragedy that has befallen their island home. I am, nonetheless, very glad to learn that all have been safely evacuated. I send my congratulations to all those who have taken part in this successful operation. Elizabeth R.

The Governor of St Helena, Ascension and Tristan da Cunha, Sir Robert Alford, was being kept informed and a decision was made to create a special issue of stamps to aid a Tristan Relief Fund. This local initiative took four values from the existing Tristan Rand-value stamps which were then overprinted with the expression 'Tristan Relief' together with an additional charge, equivalent to the face value but in sterling, which was intended to be a donation to the embryo fund. One thousand sets of the special stamps were produced. No doubt this first-known example of fund-raising on behalf of the Tristan refugees was inspired by Sir Robert who had, in a telegram the previous day to the Colonial Office, told of considerable dismay on St Helena following the news of the eruption and evacuation. However, not even knights of the realm who are also governors can interfere with the Royal Mail in its many forms. The stamps were on sale in the Jamestown Post Office for just seven days, during which 434 sets were sold, before they were summarily withdrawn and destroyed on instructions from London. In fact, the sales totalled only about £56 with a profit of some £28 for the Tristan Relief Fund, but the philatelic world quickly became excited about this 'non-issue' and speculators soon afterwards offered up to £25 for a set. How saddening that a sincere charitable effort should be halted, and yet it would go on to provide speculator collectors with a huge opportunity for personal profit far above the very modest amount donated to the new relief fund.

Meanwhile, in the Shetland Islands an emergency meeting of the Shetland Council of Social Service convened upon hearing of the plight of the Tristan refugees. Afterwards, they informed the Colonial Office in London of their interest in being a possible destination for the Tristan refugees, aware of their fishing and farming culture and the worrying de-population of their own islands. This started a debate about a new permanent home for the exiles now on their way to South Africa.

Aboard HMS *Leopard* Cdr Hicks-Beach carried out a briefing and planning

Rex Phillips' painting of HMS *Leopard* approaching Tristan da Cunha and its erupting volcano on 13th October 1961. Nightingale Island is depicted beyond Tristan to the right of the warship. The original oil painting was donated to the Tristan da Cunha Association in 2019 by Stewart Miller and in turn has been given to the island's museum. (Stewart Miller)

meeting for 'Operation Tristan' which would start the next day. Colour slides of the island were shown by Lt Cdr Wynne Edwards to give everyone an idea of the layout of the Settlement, the access to it via the Big and Little Beach landings, and the position of the new volcanic cone. On board the *Tristania* Peter Wheeler settled into his third night, no doubt joking over supper with Scotty and his officers about their break-in at his office and the fish stamps now swimming around beneath the waves.

13th and 14th October 1961: The *Leopard*'s crew set about their work

HMS *Leopard* approaches its anchorage off Tristan on 13th October 1961 with the threatening volcano behind. (Stewart Miller)

The new volcanic cone continued to grow slowly, centred on the former site of the Diamond Beacon and still no threat to the Settlement. As the *Leopard* approached Tristan during the morning of Friday 13th October its commander liaised with Peter Wheeler to agree a strict priority list for the salvage operation: personal belongings, clothing from the island shop, hospital equipment, factory equipment and valuable canteen stores. No assistance would be needed to dismantle the factory equipment as this was being carried out by crew members of the *Tristania*, and costly radio and meteorological equipment had already been disassembled and prepared for loading aboard *Leopard* on arrival. The 2225-tonne frigate reached Tristan in perfect conditions

HMS *Leopard*'s crew carrying salvaged belongings in sacks and keeping an eye on the growing volcanic cone now dominating the slopes above and east of the village. (Stewart Miller)

An ash eruption sends a toxic cloud towards the village as viewed from HMS *Leopard* on 13th October 1961. (Stewart Miller)

at 12.15pm and it was noted that the volcano was 'smoking merrily with occasional puff of sulphur'. A crew member thought the volcano resembled some Black Country slag heap emptying a column of white smoke into the air. The ship was anchored outside the kelp zone to avoid fouling its intakes and it remained in full operation to ensure that it could get away hastily in the event of a more violent volcanic eruption which could occur at any time. Wheeler and Scotty boarded *Leopard* shortly after its arrival and together with Cdr Hicks-Beach they masterminded the salvage operation in efficient military style.

Eight search parties of three crew members were assigned to various duties, all controlled by headquarters section aboard *Leopard* and in contact with those on the embarkation area at Little Beach by radio and signal lamp. Each group was provided with sacks, mail bags or mattress covers, labels, string, envelopes (for valuables and money), a notebook and pencil. Each was assigned to a cluster of cottages to search, pack up, label their contents, and mark the door using chalk or paint with the name of the owner on completion of the operation. An officer oversaw each salvage party in the first wave that Friday afternoon. One group's first target was St Mary's Church which was referred to as the 'chapel', its red corrugated roof setting it apart from the thatched houses. The team were impressed by its cleanliness, the array of flowers and feeling of peace and sanctity, heightened by the warm sunshine that streamed through the small windows. With reverence, they stripped the altar cloths and gathered up the chaplain's vestments from the vestry. After half an hour they had filled two mattress covers with clothes, chalices, books, and pictures. A radio call to HQ led to a decision to postpone moving the large pedal-operated harmonium, an organ recently presented to the island by the Queen, a task that would be undertaken later if time allowed.

One group set about clearing their first cottage. They entered through the typical sea-facing stable doors which enabled the bottom half to be shut whilst the top half remained open to the air, allowing in happier times an islander to look out to the village and ocean beyond. On opening this door, they were met with a strange sight. On the kitchen table was a teapot surrounded by five cups, some still half-filled with tea. On the paraffin stove was a saucepan of unpeeled potatoes, which would no doubt have formed the staple of an evening meal the Monday before, had not the family been compelled to leave so suddenly. More evidence of this rapid departure was seen in the tiny bedroom where a bed was unmade, and drawers left open with clothes strewn about the floor. Outside the orderliness of a pretty flower garden, inside the remnants of a hurried meal and scattered belongings. Other groups of crew members were: an HQ section of four including a 'runner'; a hospital party of three led by the ship's medical officer; a tractor party with driver accompanied by his assistant and an unloading group of four; a beach party of twelve led by the chief bosun's mate; and a survey party of three including the navigation officer and a photographer.

The tractor group drove around the Settlement, collecting the gear assembled outside the buildings and taking it to the top of the steep cliff path leading to Little Beach. Here the trailer was unloaded, and the property carried to the shore below. There it was initially placed into dinghies, which were then pulled out by line to a larger 'whaler' onto which the loads were transferred for the shuttle out to the awaiting warship. By 5pm all the cottages had been cleared and the belongings transported to the beach for transfer to the *Leopard*. The shore parties then assembled by the wooden station buildings to report their progress. In addition to the island homes, a start had been made on the expatriate homes, and clothing, linen, blankets and towels had been removed from the store.

The saddest tale was told by the dog-destruction party. The plan was to round up the dogs and shoot them at close quarters in the head as humanely as possible. However, once the first had been dispatched, the other dogs went into hiding. It was found that the .22 rifles were not powerful enough to kill a dog at any range, so the cull was postponed until the following day when marksmen with more effective .303 rifles could be landed. An officer described the immense concentration needed to keep a steady hand, and his feeling of nausea as each dog died during this terrible task - essential though it was if the island's livestock was to be protected. About twenty dogs were culled by this unlucky team on Friday the 13th. However, two puppies were spared, one of them found in Martha Rogers' cottage by Cdr Hicks-Beach during a tour of the village. In his report he was most impressed with the tremendous care his men had taken in sorting, packing, and handling the islanders' possessions. He understood that the only item that was found to have been broken when the *Leopard* was later unloaded was a single picture frame. He explained that all the crew members were profoundly sorry for the islanders whose stone cottages, church, school, community hall and shop had all been so beautifully kept. He also noted that the islanders' homes had a pitifully small number of contents compared to the 'mass of gear' contained in the more lavish expatriate accommodation.

By dusk the last items were being transported by the island's tractor to Little Beach ready for loading onto motorboats for the short journey to the *Leopard*. Back aboard his ship Cdr Hicks-Beach made his report to London and informed the editor of the *Daily Mail* that two puppies had been saved, and that they had been brought aboard and would be known as 'Tristan' and 'Cunha'. Over the next few days, they were to become the most famous (and pampered) dogs in the world. In England help from Tristan's many friends was already being mobilised. Sir Irving Gane re-opened the Tristan da Cunha Fund and contributions began to roll in from all over the United Kingdom.

On that same day Martin Holdgate, co-leader of the 1955/56 GISS and now based at the Scott Polar Research Institute in Cambridge, sent a letter to Gordon Whitefield, the official who oversaw Tristan at the Colonial Office. Holdgate effectively put in place a model for evaluating the volcanic eruption. Along with GISS geologist Roger Le Maitre, by then based at the British Museum, he offered to make available various geological specimens collected by their expedition so that the latest advances in geophysics could be used to chronicle other recent volcanic events on the island. He went on to explain that the previous volcanic eruptions on Tristan (at Stony Beach and near the Potato Patches) were small cones and lava flows; so, if the recent eruption followed their model, activity might be relatively short-lived and therefore a study could be made within a few months. He suggested that the Royal Society's Volcanology Committee should consider organising an expedition to give the Colonial Office guidance about the possible re-establishment of the weather station and the risks for resettlement. Holdgate suggested the best time for a study would be between December 1961 and March 1962 during the southern summer. Considering the rapid evolution of the emergency and the early stage of the volcanic eruption, his foresight to look ahead and plan the evaluation of a possible future for the apparently doomed community was remarkable.

Viewed from the ships and in the increasing gloom of dusk the volcano seemed to take on different proportions. A continual clatter accompanied the boulders and clinker as they broke away from its top and, after rumbling down the slope, they left a red-hot glow on the cone's north-facing seaward side. Occasionally a large lump would detach itself revealing underneath the glowing white heat of fresh lava. The crew's respect for the smouldering slagheap increased as, now back aboard HMS *Leopard*, they could go on deck to watch a continuous pyroclastic display, something akin to a major fireworks event, and pause to reflect on the dangers of the mission which would continue to accompany them the following day.

That day was Saturday 14th October and now Peter Wheeler was established aboard HMS *Leopard*. At 8am a landing was made so that the salvage operation could resume, and by late morning every cottage had been re-visited to check for any overlooked items. Only furniture remained, and doors and windows were secured against the weather. Attention then switched to clearing the belongings in the eleven expatriate

Hot cinders cascade down slopes behind and in front of the embryo volcanic cone on 13th October 1961. (Nigel Peddie)

The scene of activity at Big Beach on 14th October 1961 as work continues to transfer as many salvaged possessions as possible from the shore to the awaiting HMS *Leopard.* This view shows the fishing factory situated above the beach which within a month would be engulfed by lava. (Stewart Miller)

houses and the higher-value items in the Island Store such as wines, spirits, and beer. Hospital equipment was recovered, and all dangerous drugs were removed. A team searched for the remaining dogs and shot any that were found, though several more had escaped to the Potato Patches, and it was too late to go after them. Scotty promised that he would deal with any seen during future visits. In a symbolic gesture the naval party made sure that Union Jacks continued to fly on all the island's flagstaffs, undoubtedly aware that the unbroken British presence which had ensured that the flag had been displayed since 1816 was about to come to an end. After the ship's commander landed on Little Beach it was decided by his fellow officers and the Administrator to rename the cove 'Hicks Beach'. In the event, future maps would never record this tribute as lava was soon to bury the feature.

Re-embarkation was completed by 7pm. The islanders' homes were practically bare, and trunks and boxes of personal belongings had been stowed aboard the *Leopard.* All that was left behind in the Island Store were some foodstuffs for the use of visiting fishing crews. Items carefully loaded aboard the warship included the harmonium, and the Prince Philip Hall power plant and cinema projector which had provided weekly films to a community yet to enjoy television or reliable radio reception. The drains had been disinfected, and all buildings secured. By now the British press had picked up from a radio message from the *Tjisadane* that the Tristan children were playing boisterously all over the liner, as if unaware of their parents' plight. The *Tristania* could not move from its anchorage off the factory as there were by now about thirty boats, two motorboats and two landing lighters tied up all round it, in addition to the thirty fishing boats and three longboats held aboard. Longboats, heavy machinery and five marine diesel engines were also stowed on the *Leopard.* Scotty reckoned £30,000 worth of company equipment had been salvaged from the now virtually stripped factory, and he was most grateful for the thirty cases of beer left above Little Beach which he later said, 'kept my men happy for a few days, which they well deserved'.

The volcanic cone was appreciably active that Saturday, reaching a height of about 75 metres above the surrounding land and occupying well over a hectare of former pasture. There was still no flow of lava but there had been an almost continuous fall-out of red-hot clinker and rocks tumbling down from the summit crater and getting dangerously near to the factory, thereby threatening the tanks which still held 40,000 litres of diesel fuel. Behind the main crater several 'blow holes' were observed on the south side of the growing volcanic mass. The frigate's final report concluded that, if the present growth continued, houses at the edge of the Settlement would become imperilled within a week. HMS *Leopard* left its Tristan anchorage at 8pm intending to sail at full speed to Cape Town, with the aim of transferring the islanders' possessions to the *Stirling Castle* in time for its departure less than six days later. Fortunately, the weather had held firm, with calm conditions allowing access to the landing beaches. As the ship sailed away, the crew were able to see not only snow on the top of Queen Mary's Peak but also terrifying flames above the Settlement 2000 metres below. The timing of the entire evacuation mission had been immaculate; within hours red-hot lava flowed from the new cone for the first time.

Sunday 15th October 1961: Grim fascination at Tristan, generosity elsewhere

Overnight Scotty and his *Tristania* crew watched fascinated as brightly glowing rocks cascaded down the flanks of the new volcanic cone directly behind the factory. He considered its site above Big Beach too dangerous to approach, so he brought aboard as many island fishing dinghies as possible and arranged those remaining to be towed behind. The ship then headed for Inaccessible Island where it anchored off Salt Beach. Here the crew built a shed to act as a temporary storage depot using salvaged timber. Various items of fishing gear were brought ashore, as were the dinghies, which were hauled above the storm beach and stored upside down in the open. Scotty knew that the boats would rot but he hoped to return and build a boat shelter when more timber was available. The fishing boats *Tristania* and *Frances Repetto* normally fished the outer islands of Inaccessible, Nightingale and Gough, leaving Tristan itself to the shore-based islanders. Often, up to six island men joined *Tristania*'s boat-based fishing operation, but now, of course, this support was no longer available. Rather more significantly, the Development Company was facing an overwhelming crisis: its valuable Big Beach factory appeared to be the first building likely to succumb to the merciless lava.

With the dawn of Sunday 15th October, the Tristan islanders had been five nights aboard the *Tjisadane*, and as they gathered for a service conducted by Father Jewell, they knew that Table Mountain would be in sight the following morning. Undoubtedly many private prayers were offered up for their future. Martha Rogers, the island's Head Woman, said, 'There is God's hand in this – we were meant to leave, and He will look after us.' Meanwhile, aboard the Royal Navy frigate plans were underway to establish a *Leopard* Tristan Relief Fund to provide spending money for the islanders during their voyage to Southampton. Tickets were sold for a raffle at a cost of 5s (25p) each with donated prizes that included a transistor radio, a £5 note, and three-months' free haircuts from the ship's barber who went under the nickname of Buffer. Collecting boxes were placed around the ship at the entry to film shows, and any profits from the daily tombola sessions were added to the mounting sum. Also, the small collection of mail found during the salvage operation in the Administration Office was carefully examined and each item was stamped with a 'Commanding Officer – HMS *Leopard*' cachet and dated 15th October 1961. Later sent on to the UK where these covers received the Maritime Mail postmark, they would soon create great interest amongst philatelic enthusiasts who recognised that they could well turn out to be the last mail ever to come from Tristan da Cunha.

16th-20th October 1961: interlude in South Africa

The Dutch liner *Tjisadane* was a regular visitor to Cape Town, but its arrival filled with Tristan refugees into Table Bay on Monday 16th October was like no previous approach. A flotilla of boats, including a police launch, sailed out past Robben Island to greet the evacuees who by now had become the focus of the world's press. The local *Cape Argus* newspaper planned a media coup by sending out at noon the motor yacht *Chycaron* with four of its reporters on board. One of the *Argus* men used a loudhailer to hold a noisy conversation with Rev. Jewell, hoping to sign him up for a scoop of some pictures and a story. Once the *Tjisadane* was anchored outside the dock and even before it began proceeding to the quay, many more boatloads of newsmen closed in and started bargaining for photos and stories. One of the ship's officers was seen with a handful of photos, whereupon an *Argus* reporter leaned over, grabbed them, passed a handful of cash into the man's hand, and the boat left to rush its prize to the news office for the next day's front page. The liner then moved towards its berth and those on board thronged the decks, stunned at the reception on the dockside where thousands of well-wishers were assembled. Inevitably the mood amongst Tristan islanders was sombre as it was assumed that this was the end of life as they had known it. Rev. Jewell summed up their position soon after coming ashore: 'They know they can't go back. The volcano was a pretty big affair when we left and it's ten times bigger now. They know the island is written off from the living point of view.'

Speaking to one of the reporters Sidney Glass said that they were all very unhappy about leaving but added that it would be better going to Scotland (where his great-great-grandfather had come from) than staying on a volcanic island. He also praised Peter Wheeler with the words, 'He is a really good man, the way he organised things. He done it very nice.' That evening the people from Tristan settled down for their final night aboard the Dutch liner. The following day the *Stirling Castle* arrived from Port Elizabeth and docked opposite *Tjisadane.* Before disembarking, many grateful passengers sought out the ship's captain to thank him for rescuing them and bringing them safely to the mainland. Sidney was reported as saying, 'Thank you captain, thank you, and may God be with you always.' Others came in shyly, shook his hand and murmured their appreciation. Capt. Giel was apparently deeply moved by their response, perhaps because he was used to a lack of thanks from many of his more fortunate passengers. A short while after, the islanders and their meagre belongings were transported by bus and lorry around the dock to board the Union Castle liner which was scheduled to set sail three days later. Their hasty departure from Tristan was evident from the random prized possessions they had brought with them, some carrying hurricane lamps, others treasured pictures. Presumably, Dennis Green's three oranges had already been enjoyed by then. All the refugees were now being closely scrutinised, and some were interviewed again by the press.

As the islanders filed silently aboard the massive liner they were met by smiling and

Islanders gather on the deck of the *Tjisadane* on 16th October 1961 as the ship arrives in Cape Town. (Tristan Photo Portfolio)

Part of the congregation in St George's Cathedral, Cape Town on 17th October 1961 during a Holy Communion service specially organised for the Tristan islanders to commemorate their safe arrival in South Africa the previous day. (Peter Wheeler)

friendly stewards who took them to their cabins. They were issued with large, lavender-coloured tourist tickets, not that they needed identification as they had become celebrities, easily recognisable by their swarthy appearance and their mode of dress. Later the islanders stepped back on shore in Cape Town's typically bright sunshine to travel by special buses to St George's Cathedral to give thanks for their deliverance from the catastrophe that had forced their rapid departure. Virtually the whole community attended: grey bearded elders, women with their distinctive headscarves, men in Sunday-best suits, all of which rendered them highly conspicuous in the relaxed atmosphere of the Cape. Fathers led small children while others carried babes in arms, all scrutinised intently by a fascinated public and an army of photographers. The cathedral service was conducted by the Assistant Bishop of Cape Town Rt. Rev. Roy Cowdry who had visited Tristan in 1959 and was therefore already known to the people. It was a moving ceremony, starting with *God Save the Queen* and followed by the hymn, 'Lead us, Heavenly Father lead us, Oe'r the world's tempestuous sea.' Many older islanders were overcome with emotion. Onlookers, including seasoned reporters, were moved by the homely atmosphere, how well-behaved the children were, and how babies were handed from one to another, so that the adults could all take it in turns to approach the communion rail alone. Bishop Cowdry gave a short address, saying that God helped them in their time of crisis and that they must never let Him go. Many regarded their escape as a miracle and that God meant them to escape unscathed from the volcano. Few who witnessed that Cape Town service would argue against that view.

After the event, the islanders were entertained to a buffet lunch of tea, scones, sandwiches and cakes offered by the women connected with the cathedral. The evacuees appeared shy and only a few of them spoke, but they all laughed happily when Bishop Cowdry lifted his cassock to show that he was wearing Tristan socks in their honour. The islanders were also to learn that day that the media would be eagerly watching their every move and attempting to get interviews when they preferred not to talk and were eager to preserve their privacy. But there were many friends around who were keen to renew their links with the islanders. Allan Crawford, now living in Cape Town as Port Meteorological Officer, was regarded as the island's welfare officer, and he described it as one of the most impressive services he had attended. A parallel ceremony, delayed by an hour to enable the islanders to attend Confession, was arranged for the 24-strong Roman Catholic community in St Mary's Cathedral nearby. That group included Agnes Glass who, as Agnes Smith, had married Joseph Glass in Cape Town some sixty years earlier; in 1958, as a mark of acknowledgement from her church, she had been awarded the prestigious Benemerenti medal for her tireless work in creating and maintaining the world's remotest Roman Catholic community. This was the first time the Roman Catholic congregation had ever gathered together in a church, since the tiny group of the Glass and Rogers families normally met for Mass every Sunday in a room kept as a chapel in 'Aunt Aggy's' cottage where Agnes or her son Cyril would conduct the prayers.

Some islanders already had strong links with South Africa. Over the coming days Ada Green was invited out for trips by her local pen friend, and with her husband Dennis and two-year-old daughter Valerie was taken to a drive-in cinema, a form of entertainment as popular then in Cape Town as in the USA. Thirteen-year-old Timothy Green's older pen pal took him for a ride in a sports car, up Table Mountain, around an aquarium, and bought Tim his first ever ice cream which he described as 'frozen milk'. Elsewhere during their stay, African Consolidated Theatres arranged a special entertainment, with the venue and actors waiving their normal fee to raise proceeds for the islanders, whilst fifty members of the group visited the Cape Horticultural Society's flower show in City Hall. Two of the refugees made enquiries about remaining in South Africa: a young girl who had a chance to join a family in Kimberley, and a

30-year-old senior Tristan factory employee who considered living in Cape Town. They were advised by Peter Lewis. the British Embassy official, to stay with their families and continue to England, where they would be looked after by the UK Government. Without passports, or the appropriate documents now required in South Africa to authenticate their racial origin, they decided to head for Britain and so the community remained together.

Alfred Green was taken with Willie Repetto to Fish Hoek to meet 90-year-old George Cotton, a man who had migrated from Tristan to South Africa in 1890 at the age of 20 and who was Alfred's godfather. Prior to that, Alfred had his first experience of speaking on a phone. When he heard the voice on the other end of the line ask, 'Is that you Mr Green', he was so startled that he dropped the handset. Later, when he had recovered from the shock, he told a newspaper reporter of his astonishment at the lights of Cape Town and the experience of his first-ever drive in a motorcar.

Hardly any of the 264 islanders now accommodated in tourist-class cabins aboard *Stirling Castle* had ever been away from Tristan. They were aboard a ship with enough capacity for 784 passengers, 246 of whom were in first-class accommodation. The ship, over 200 metres long and weighing more than 25,000 tonnes, significantly dwarfed the *Tjisadane*, but in tonnage it was forty times bigger than the *Tristania* and a staggering eighty times that of the tiny *Frances Repetto*. The islanders were impressed with the dining room where they enjoyed waiter service, white linen tablecloths and excellent meals. During the course of the day HMS *Leopard* arrived in Cape Town and soon various officials had made their way aboard. Peter Wheeler and Cdr Hicks-Beach held a press conference in the wardroom after which the reporters were served with drinks. This proved extremely popular with the locals as it was Election Day in South Africa and all the bars ashore were consequently closed.

Whilst this was happening Scotty paid his last visit to the fish-canning factory on Tristan, making a quick trip ashore to 'grab' some oxygen and CFC gas bottles that had been left behind. Rocks from the erupting volcano were now coming right down onto Big Beach, so undeniably this was a dangerous mission for him to complete. Back in Cape Town all the gear had been unloaded from HMS *Leopard* by 7.15pm, and by midnight all the personal property was laid out for identification and transfer to *Stirling Castle* next morning. As there were only seven island surnames, there were corresponding areas for each family, and the careful labelling process carried out by the *Leopard*'s crew on the island proved its worth. Rev. Jewell helped confirm the ownership of the few unidentified parcels.

Peter Wheeler and his family were re-united and stayed with friends in the Rosebank suburb of Cape Town. He remembered jumping every time a lorry went down the road, the vibration reminding him of the tremors that had shaken the Tristan Residency just a few days before. Around the corner lived Allan Crawford who entertained his close Tristan friends Arthur and Martha Rogers, with whom he had always stayed during his visits over the years. Arthur was philosophical about their fate believing that their safe evacuation was God's purpose, adding, 'This would not happen without it was His will.'

On 19th October, a working party from HMS *Leopard* assisted in the safe transfer of the islanders' personal possessions to the *Stirling Castle.* The Union Castle Line had already filled almost every available space on its ship with cargo and their officials were therefore concerned that insufficient room could be found for the additional luggage. In the event they were able to receive everything, including three island longboats. At midday the Chief Islander Willie Repetto, Head Woman Martha Rogers and Rev. Jack Jewell went aboard HMS *Leopard* for a ceremony during which Cdr Hicks-Beach handed over the £235 raised for the ship's Tristan relief fund and a heraldic crest from the frigate. Also returned were a similar plaque from HMS *Protector* and a Union Jack which had flown from a staff outside the Prince Philip Hall. Back in Britain the London-based SPG who funded and organised priests for Tristan announced that they would be donating £1000 for the islanders. That evening, following Alfred's Green visit earlier in the week, three buses carried most of the islanders to Simonstown for a party held in the Royal Alfred Hall which was sponsored by the local council. Bishop Colin Winter welcomed the islanders and blessed the meal. The refugees were entertained to a feast and danced to two bands. Darryl Felix, who was present as a youngster commented, 'I did not understand a word the evacuees spoke.'

The following day was to be the last of the Tristan islanders' stay in Cape Town, and they had impressed all who had met them with their soft-spoken manner and quiet demeanour. Their Anglican and Roman Catholic congregations had received two large bibles, whilst local people had also generously donated clothing, toiletries and school equipment, and the South African immigration authorities waived the usual immigration fees. Peter Wheeler expressed his gratitude to Cape Town's mayor Mr Honikman for the help his city had provided and formally handed over a Tristan longboat as a gesture of thanks from the island community. A successful cash appeal enabled the Mayor to present a set of envelopes to every islander containing 50s (£2.50) for adults and 30s (£1.50) for children, all accompanied by a hand-signed card. The Mayor and Bishop Cowdry bade the Tristan refugees farewell and just before the gangways were lowered Chief Willie Repetto issued a parting message to the city which was published the following day: 'The people of Cape Town have been more than kind to us. We never expected such hospitality and generosity. Good-bye South Africa, now we are looking forward to England.'

The *Stirling Castle* departed Duncan Dock at 4pm, waved off by a crowd of over a thousand who gathered on the quays despite a light drizzle. Coloured streamers were thrown up towards the islanders who lined the vessel's decks as they started the next stage of their exodus into an uncertain future. Joining the evacuees, though accommodated in the first- class section, were the expatriates who had travelled with them aboard the *Tjisadane*, including Margaret Wheeler and her three children. Not their father, however. He remained in Cape Town poised to fly ahead to England to prepare for the refugees' arrival. Speaking later of the islanders' time in South Africa, Wheeler thought that the younger people had enjoyed it, but those over forty had been bewildered by the cranes, cars and traffic lights. All were astonished by the shops. He added that the generosity of Cape Town had been 'quite incredible'.

Late October 1961: Preparing for the refugees' arrival; Lava destroys the factory

At Tristan on 21st October Scotty returned to the island and found lava had reached Big Beach and molten rock was nearing the factory. He estimated the cone to be almost 120 metres high and noted that a stream of lava was closing in on Big Beach and the fishing factory. At the same time, as the islanders and expatriates settled into their long journey north, they became aware that Union Castle company policy prevented the Tristan doctor Norman Samuels from continuing to provide medical treatment, since this had to be carried out by the ship's own surgeon Dr Flew. In the end a more even-handed arrangement was made for them to work together, and any treatment given would be provided at the British Government's expense.

At Johannesburg Airport during the evening of 22nd October Peter Wheeler boarded a BOAC Comet aircraft bound for Britain. At 11am the following day he landed at London Airport (only in 1966 would it be renamed 'Heathrow') to be confronted by the media who were keen to know more about the Tristan volcano and the future for the islanders. His lack of suitcases proved a source of interest and one article published the following day started with the line, 'A man from the Colonial Office flew home to London yesterday with only a shirt as luggage after six months in the loneliest place on earth.' In an interview he said, 'The island is dead, finished. I do not think anyone will want to live there again.' Asked about the future for a people who in the main had never seen cars, trains, or planes, he added, 'The older ones are completely bewildered; the youngsters are anxious to sample the bright lights they have never seen.' Even now there was speculation about where the Tristan refugees might settle, and an early expression of interest had already been received from the Shetland Islands. Asked in that context about the islanders' future he replied, 'I think it would be wrong to try and create an artificial Tristan for these people. Most of the men have skill as fishermen and boatmen and some will be assimilated into the life of Britain. It has been a saddening experience. The main thing is nobody even got wet during the evacuation.'

The following day, after a brief visit to his home in the New Forest, Wheeler moved into a London hotel to be able to concentrate on preparing for the arrival of the islanders. His changed role meant he would no longer be taking the lead in making decisions in the way he had done so decisively when organising the safe evacuation from the island. As a result, he played no part in the early plan for the refugees to go into temporary accommodation at the disused Pendell Army Camp in Surrey, immediately after their arrival in Southampton. The site was chosen primarily because it was available and also for its proximity to Whitehall, rather than for any comfort. The former barracks had housed a searchlight unit during the Second World War, and an Anti-Aircraft Command operations room built in 1951 had remained in use until 1960. This was a fortunate time to find a place available at short notice for all the islanders since many service camps were being de-commissioned. Furthermore, a shortage of labour was encouraging large-scale immigration to Britain to ensure there were enough workers to staff the factories, farms, and service industries of the country. In that respect, the modest influx of English-speaking Tristan refugees, used to hard work and with a strong community ethic, was generally welcomed. Wheeler accordingly liaised with Lady Reading, head of the Women's Voluntary Service (WVS), who he described as 'absolutely marvellous' as she mobilised Surrey WVS members to help prepare the empty camp for the evacuees' arrival.

On Friday 27th October Scotty sailed again to Tristan and dropped off some dinghies on the east side of the island at Sandy Point. When he steamed around Big Point, he had a considerable shock as the factory had been completely obliterated by lava which was now extending 100 metres out to sea towards the anchorage where it was steaming vigorously. It seemed to have split in the middle to form a kind of gulch where red-hot lava appeared to have poured out to cause a huge explosion. As a result of this violent occurrence, the three fuel tanks at the Big Beach factory holding 40,000 litres of diesel will have ignited, thereby producing a blast eastwards during the previous few days.

On 30th October, a hectic week of preparations began at Pendell Camp to ensure that the accommodation would be ready for the Tristan islanders' arrival four days later. The camp had been closed for months and the huts needed to be transformed into a reasonably acceptable living arrangement until a more permanent home could be found for its temporary occupants. Accordingly, staff from the Colonial Office, Administrator Peter Wheeler, and a squad of soldiers were soon fully occupied in this dramatic race against time. Supporting this enterprise was a group of 14- and 15-year-old pupils from St. Catherine's School, Bletchingley together with two of their teachers, and volunteers from the British Red Cross and the WVS.

Back in the South Atlantic that week Scotty took a final look at the Tristan volcano before returning to Gough Island to complete the spring fishing trip. The lava flow was now 800 metres wide though it had avoided the village houses. The lava front was creeping seawards and by then about 400 metres north of the buried factory site where it seemed, in Scotty's words, to be 'rising out of the sea'.

Photo believed to be taken by Capt. Scott from aboard MV *Tristania* on 27th October 1961 which is the first known to show lava erupting from the new volcanic cone. The front of the lava, seen here producing steam as it reaches the sea, has completely engulfed the Big Beach fishing factory, presumably accompanied by an explosion as the diesel fuel stored there ignited. (Peter Wheeler)

On 31st October Wheeler sent a telegram to Father Jewell who oversaw arrangements for the evacuees aboard the *Stirling Castle*. He was able to pass on the information that they would be staying at Pendell Camp near Redhill whilst their resettlement was under discussion. He explained that the situation was a temporary arrangement and that the facilities were communal, unlike the hotel standard accommodation they were enjoying on the ship. Islanders would be required to cook and clean but would be helped by volunteers from the Red Cross and the WVS. Wheeler outlined various staff arrangements, envisaging that Rev. Jewell would be in charge of all social and welfare issues. A school would be run by Miss Bennett and Miss Downer, and a sick bay run by Dr Samuels and a qualified nurse. Finally, Wheeler advised Jewell to be prepared for the members of the press who planned to join the *Stirling Castle* in the Canaries for its final leg to Britain.

Silver service for the Tristan refugees in MS *Stirling Castle*'s dining room. (British Pathé)

On 1st November reporters and photographers visited the camp in Surrey to observe how the preparations were progressing. One report described a hive of marvellously organised confusion, adding that above all the terrifying helter-skelter Peter Wheeler, the tall, dark, cool man from the Colonial Office, could be found sewing threads of order into all the energy being spent so willingly. The WVS were described as 'quite unstoppable, doing a great job, keeping the tea flowing, scrubbing down the tables in the cookhouse and in general working like Trojans'. The St. Catherine's School pupils were seen rushing back and forth across the parade ground carrying blankets and pillows, seemingly enjoying themselves, perhaps because they were missing their usual classes.

The visitors were shown the living accommodation, which one of the NCOs described as 'billets'. The army had supplied oil stoves for heating and the long huts were partitioned-off by screens to separate the various families, each identified by the names that appeared on the doors. Some name labels were lost in translation and included 'Joe Repekko' and 'Willy Lavarello', a situation that no doubt amused the islanders when they arrived 48 hours later! Two pianos, a billiards table, a selection of books and a TV were provided to afford some interest and entertainment for the islanders. The press was informed that the people they had newly named 'Tristans' would be tired when they arrived. Nothing but food and sleep was planned for the first few days; parties and dancing would come later.

Aboard the *Stirling Castle* an *Evening Standard* reporter highlighted the sparse possessions of the islanders, pointing out that some had no more than '£1 8s 7d [£1.43] each and some clothing'. Nevertheless, they did not seem concerned by a lack of money. A further £370, including a £120 collection made by the officers and crew on the *Stirling Castle*, was to be shared amongst them. Alfred Green was quoted as saying, 'As long as I'm still living, I don't worry.' As he observed the men folk helping the older women and young mothers with the babies, a journalist wrote how impressed he was with the great calm of the islanders in their predicament. He therefore agreed with Rev. Jewell when the priest told him that the islanders 'may be able to show people in England something valuable about what real community life can be like'. Peter Wheeler, speaking to reporters back in Britain at around the same moment, thought the islanders would find it a strange and confusing experience, having moved from their remote island, undertaken a trip on a luxury liner, and soon to be obliged to settle down in a cold army camp. He was conscious that the adults would have to cram into a few months what locals had spent twenty years in learning, road safety being a case in point. He was also keenly aware that newly arrived islanders would have no natural resistance to germs and that therefore the camp's sick bay would prove to be an essential facility.

On 2nd November, the camp's commandant and his men moved out of Pendell Camp and left the final arrangements to the team of Colonial Office staff, the WVS and school pupils. By then the sick bay had been set up and would be led by Vivien Woolley, a resident trained nurse from Derbyshire. Twenty years before she had accompanied her husband Surgeon-Cdr E Woolley to Tristan where he had led the war-time naval station. The Surrey Branch of the Red Cross provided all the equipment and some of the staff, and Mrs Woolley was to be further assisted by auxiliary support from the local branch of the St John's Ambulance Brigade.

Islanders take in their first sight of England as they arrive in Southampton aboard MS *Stirling Castle* on 3rd November 1961. (British Pathé)

Edith Repetto follows her husband Arthur (carrying her handbag) as they step ashore on the quayside at Southampton on 3rd November. (British Pathé)

Friday 3rd November 1961: the refugees arrive in Britain

In the darkness shortly before dawn on 3rd November the *Stirling Castle* sailed up Southampton Water and a few of the islanders gathered at the rail as the liner slid slowly to its berth. From 6am the ship was invaded by politicians, dignitaries, organisers, reporters, and cameramen. The official party was led by Rt Hon. Hugh Fraser MP, Under-Secretary of State at the Colonial Office, his colleague Gordon Whitefield, and Peter Wheeler who was re-united with his family. Also allowed aboard to join their families were three islanders who had been living in Britain: 15-year-old Valerie Glass, who had immigrated to England in 1952 at the age of 6 under the guardianship of the teacher Mrs Handley; 18-year-old Jennifer Rogers; and 20-year-old Sarah Swain. Fraser addressed the islanders and told them that the UK Government willingly accepted responsibility for looking after them and that plans for their future should not be rushed. He was listened to in silence and generously applauded. After breakfast, a press conference was held, attended by over sixty journalists and television staff. Questioned about the islanders' future Chief Willie Repetto said,

We may have to stay here but most of us want to return. You see it is the place where all of us were born. We spent our lives there and it is the only life we know. None of us have ever seen England before, although one or two have been to Cape Town.

But on that sobering November morning, with the only news from Tristan indicating that the volcano continued to threaten the village, it seemed inevitable that the islanders' future was now in Britain.

Just before 9am the islanders walked down the gangway onto the dockside where, according to *The Times*, they received a welcome that was 'warm, emotional and unrestrained'. Many friends and other relations had gathered here, including the former Administrator Peter Day, now working as a personnel officer in a Leicester business. These reunions remained private affairs but Rev. Philip Lindsay, who as a young man had been a lay preacher on Tristan between 1927 and 1930, described his experience in the following terms for the *Leicester Mercury*:

No words of mine can ever express those first few moments as I went back in time to the days when we shared the joys and hardships of the days when things were not very plentiful. I knew them and they knew me at once. I was able to put a name to a face and tell them their age, and all the things that we used to do, whether it was turning the sheep down from the mountainside or looking for something to eat from the sea. To see those faces beaming with delight, to see the look of almost unbelievable wonder at the thought that must have been in their minds, that you had really taken the trouble to think about them, beggars description. As I spoke to these kindly people I wondered what was uppermost in their minds, and maybe it was summed up in the words of Mary Swain when she said to me, 'It has been a terrible blow to us to have to leave, but God is good, and we must leave it in His hands.' What a wonderful faith when you have lost your ALL; your home, your simple livelihood, the very ground that you have known since the day that you were born.

The islanders soon boarded a fleet of coaches and embarked on the final leg of their enforced departure from their homes, one that took some three-and-a-half hours because a stop for tea had been organised and a nearby store was crowded as eager islanders bought the owner's entire stock of eating apples in their first opportunity to shop in England. The journey was followed by the press and reported across the world. Thus, the citizens in the American mid-west state of Ohio read in the *Toledo Blade* this quote from Harold Swain as he described that coach trip: 'I have never travelled so fast in my life. I am told that our six buses were moving at times at 60 miles to the hour [100kph]. It is hard to believe.' With copy like this eager reporters sought out similar quotes from the islanders who soon became aware that the media's interest in them was not always positive.

At Pendell Camp a large group of people had gathered to welcome the party's arrival, amongst them some two hundred women of the Godstone and Caterham division of the WVS. The smiling, swarthy faces of the 'Tristans' (as they were now called by the press) peered through the windows of the coaches as they swept onto the former parade ground and their eyes lit up with recognition again and again as urgent waves and shouts attracted their attention from many old friends. As each coach drew to a halt a WVS volunteer climbed aboard, welcoming the passengers to Pendell, and taking them to their huts. Here they were shown around where they were going to live and escorted to lunch in the dining room. The WVS noted that many islanders had been unwell on the coach journey and were apprehensive and tired. The *Surrey Mirror* reported that:

BBC cameras whirled in the hurly-burly of hugs and kisses and the old-time English accents contrasted with the smart voices of clergymen and government officials. Names on both sides were instantly remembered as people suddenly found themselves confronted with friends they had not seen for years. Dressed in neat old-fashioned clothes, the Tristans were led away to their accommodation. They had arrived in their new home and all the world was looking in.

The entire consignment of the islanders' possessions had by then been unloaded from the *Stirling Castle* and later a convoy of just two trucks brought it to Pendell, underlining the meagre extent of their belongings. Hoping to provide a familiar first dinner the WVS team prepared fish pie with mashed potatoes and cauliflower, followed by stewed apples and custard. Afterwards islanders gathered round a television set during their first evening and watched the news which led with a report about the birth of a son to Princess Margaret that same day. Later, the various families retired to their own accommodation, sharing communal bathrooms and separated from the next family only by curtains, but nevertheless brightened by flower arrangements generously provided in a community gesture by a local florist.

Arrival at Pendell Camp, Surrey on 3rd November 1961. Those in front are the Rogers family with father Rudolph carrying Patrick, whose older sister Doreen and her mother Marjorie follow behind, while Ronald, Alfred and Roy step out in front. (Tristan Photo Portfolio)

Part 4:
Life in Surrey and Hampshire
1961-1962

Settling into a chilly Surrey army camp

The islanders' first full morning in England dawned cold and clear with a biting north-west wind blowing towards the huddle of wooden army huts in which they were now living. They were familiar with the similar sort of constructions that made up the old naval station back home, but these had never been slept in by any of the families who had always been snug in their thatched stone cottages and had never experienced frost. Families, sleeping on iron beds, were only separated by curtains from those who shared the large huts. As well as island families, those of expats such as Peter Wheeler and Rev Jack Jewell also endured this sparse communal living. Young Simon Wheeler's abiding memory was one of the disappointment he felt that onion soup was served up time and again.

A reporter commented that only the children seemed really happy as they explored their new world. All morning a small boy was seen sitting astride a bicycle, pushing himself down a concrete path beside his hut and by lunchtime he had his feet on the pedals and had learned how to ride. Outside the huts, women in overcoats and hand-knitted woollen stockings stood talking in groups. Challenging a member of the press Aunty Martha Rogers said, 'You will not find us like this at home. We'd be about our work and the men out fishing or in the fields. Idleness is not for us. Some are lucky. They have work helping in the kitchen.' Many went shopping in nearby Merstham, able to spend some of the cash given them on board the *Stirling Castle* and an allowance initially provided at Pendell Camp to tide them along.

A large room in the former NAAFI building had been allocated as a playroom for the children and kitted out with toys, books, puzzles, records, prams and tricycles. An unwanted piece of red carpet had been found to cover the floor, and brightly coloured pictures had been put up to decorate the room. Willie Repetto and two of the other island men were invited to watch a football match that afternoon between Redhill and Wealdstone. They were welcomed by the club's officials and after the game were presented with two footballs to take back to the camp.

On Monday 6th November, the camp school opened in the former sergeants' mess, by now transformed into a pair of classrooms. It was staffed by Rhoda Downer, who had previously taught in the island school, and Ethel Bennett, who was their current teacher and who had continued her work aboard the *Stirling Castle*. Lessons progressed, often watched in the early days by the press. One newspaper published a photo showing 13-year-old Barbara Swain painting a watercolour of the volcanic eruption that had so dramatically threatened her home. Press stories reported the Tristans' bewilderment with metalled roads, fast cars, traffic lights and modern shops. In fact, the refugees were adapting remarkably quickly; they proved to be prudent with their money and were careful not to wander far until they got used to the local road network. On the camp itself the WVS ran a small shop and an information bureau. Profits from the shop were used to buy materials for the islanders, such as wool so that the women could continue knitting. One of those volunteer shop assistants said, 'They have caught on to our money system quickly – at least some of them. The rest pull out a handful and leave you to take what you want.' Shopkeepers in Redhill and Merstham found the islanders quiet and polite, 'but a bit difficult to understand'. A local newsagent set up a stand at the camp to sell papers every day so islanders could now enjoy the novelty of reading up-to-date news for the first time in their lives.

By Wednesday 8th November, the cold Surrey weather was beginning to take its toll and, for the first time in his life, 42-year-old Cyril Rogers was confined to a bed with acute bronchitis. The WVS volunteers distributed further clothing to the islanders, including suits that the men would need to attend job interviews. Especially in demand were children's coats, wellington boots and warm underwear, as well as dressing gowns and slippers for those in the sick bay.

Many offers of help came in, one notably being made by a nearby hairdressing school which offered to give demonstrations and free hair styling for the island women. The dryers and stylish models arrived; the island women were encouraged to try hair tints of pink, gold, and red, but few took advantage, so the scheme only lasted a fortnight. Dancing classes proved more popular, and the WVS arranged for a trained teacher and some of their younger WVS members to give the islanders various lessons in the trendier rock 'n' roll dances as well as 'the twist'. With both a gramophone and an accordion the Tristans were eager to join in and re-started Saturday night dances, in Surrey instead of the Prince Philip Hall.

On their first Saturday in Surrey, 47 Tristan children were taken on a coach trip to the Odeon in Redhill to join the highly popular morning children's film club. The young islanders had never heard a noise as loud as that from the packed young crowd. 18-year-old Ches Lavarello was quoted as saying, 'These people here make more noise in a minute than we do in an hour,' and laughed even more when the English children gave three hearty cheers for the visitors from Tristan. A special open-air road-safety exhibition and demonstration organised by Reigate Road Safety Committee outside the cinema had to be cancelled as it was pouring with rain. Instead, the Tristan children were given a copy of the Highway Code and local instructors took cycles onto the cinema stage and answered a few questions on bicycle maintenance put to them by a police sergeant.

When teacher Ethel Bennett was asked about how the children were getting on, she replied that they were 'so overwhelmed that they do not think to ask questions. They just accept.' The Girl Guide Commissioner for Surrey came to the camp and soon many of the Tristan girls were enrolled as guides or brownies. Senior guides visited on Saturday mornings to join in various games with the girls, and a cub leader played football and hockey with the boys.

Tristan Scouts attending a meeting of the 22nd Purley Scouts while they were at Pendell Camp in November 1961. The boy to the left without uniform is Ken Green and other islanders are, in the middle row from the left, Frank Rogers, Douglas Swain, Stanley Swain, David Swain, Timothy Green and in the middle of the front row Ronald Rogers and Barney Swain. (Terry Farebrother)

The Surrey army camp was always going to be only a temporary home lasting just a few weeks, and yet at the beginning of their second week in England forty islanders had taken up paid jobs with local companies. Eleven were taken on at the Monotype Corporation's type-manufacturing plant at Salfords, two at the Fuller's Earth Union Copyhold Works at Redhill, six at the firm of Messrs Nettings, three at the Nutfield Manufacturing Co., ten by a firm of industrial caterers at Pendell Camp, whilst four joined the ground staff at Reigate Heath Golf Club and four at the Standard Brick Co. at Redhill. Allan Green was one of those at the golf club and he was disappointed that on his first day bad weather had kept golfers off the course as he was keen to watch a game. His work involved sieving soil (always called *mould* by islanders), which he enjoyed despite the cold and rain, and he expected to deposit most of his week's wages of £9 in the bank. With his three colleagues he caught the bus home on their first night, but they lost their way in Merstham and spent over an hour wandering down deserted country lanes before they were put on the right road again. Harold Green spoke of his first day's work at the Redhill brick company where he worked from 7am to 5.50pm. He had enjoyed stacking bricks, but was more pleased by the offer of a free lunch in the works canteen for the first fortnight. He planned to save most of his £15-a-week wages and hoped to buy some crockery for his wife Amy. Harold was grateful for the warm welcome he had received at the brickworks from the other employees, who made him feel quite at home and were 'jolly good chaps'.

Photograph at Pendell Camp in November 1961 of the Tristan islanders and the expatriates staying with them in and assembled alongside one of their longboats. (Peter Wheeler)

16-year-old Gerald Repetto wrote a letter to Allan Crawford explaining that the islanders had settled down nicely at Pendell Camp. He was about to start factory work with Roger Glass earning £4 14s (£4.70) a week but was aware that much higher wages were being paid to older men. Gerald noted the chilly, wet weather, and the fact that most of the people were suffering with colds and some, including Aunty Martha, were in the camp's sick bay. He remembered the kindness of the people in Cape Town, the pleasure of the trip on the *Stirling Castle*, and the big welcome they had received in England. He was delighted to see friends who had lived on Tristan and had stayed with Martin Thompson, had ridden nearly 8km on his bicycle, and was getting used to the traffic and the road signs. He also enjoyed the camp social life with its television, billiards, darts, and table tennis, as well as the dancing.

Photograph showing the scene in front of the Royal Exchange in the City of London on 16th November 1961 to launch a *Daily Telegraph* appeal to support the Tristan refugees. In the longboat Carlisle are two island men, together with the Lord Mayor of London, Sir Frederick Hoare, and Lady Hoare. Bearded longboat coxswain 'Big' Gordon Glass is behind Sir Frederick and, wearing a bowler hat behind Lady Hoare, is Sir Irving Gane, then Chamberlain of the City of London. (Peter Wheeler)

On 16th November the *Daily Telegraph* arranged for the 10-metre Tristan longboat *Carlisle* to be transported from Pendell Camp to stand in front of the Royal Exchange in the heart of the City of London, just across the road from the Bank of England. Here the newspaper launched an appeal to raise £50,000 to help equip the new homes that the Tristans were expected to move into when they vacated their temporary accommodation in Surrey. The event formed the public re-launch of the Tristan da Cunha Fund by Sir Irving Gane, who was now in the eminent position of Chamberlain of the City of London and therefore ideally placed to organise a high-profile campaign. The Lord Mayor of London, Sir Frederick Hoare, presented a cheque, a longboat crew of islanders, led by coxswain 'Big' Gordon Glass, manned their oars, and Sir Frederick and Lady Hoare joined Sir Irving on the boat for a photo-shoot. What an incongruous contrast in appearance they all made: the male dignitaries in their typical City-style black bowler hats and pin-striped suits, by the side of the rugged island men with their clothing of assorted oilskins and flat caps. Over £4000 was raised on the first day alone. A later newspaper report mentioned that the longboat standing on the pavement alongside the famous Threadneedle Street was a 'reminder of the responsibility we all share for the islanders' future. They are refugees in the most absolute sense, wholly dependent on us not only for present succour but for permanent resettlement.'

Tristan women at Pendell Camp had by then taken complete control of the washing-up and the cleaning of the huts and had started to share in the preparation of meals, all of which were produced in the camp kitchen and eaten in the shared dining room. A *Surrey Mirror* article was led by the headline, 'Too cold for short skirts, say Tristans', before quoting one of the Tristan ladies, 'We think the women here are beautiful, but we can't understand why they are wearing their skirts so short. Don't their legs get cold without any knitted stockings?' The journalist found the evacuees eager to learn all they could about dress and make-up. Nylons had quickly replaced woollen socks for the younger islanders and local shopkeepers said the island women were far easier to serve than many of the locals, being so polite and eager to do the right thing. Four young Tristan women found retail jobs at Woolworths in Redhill, and many of them were heard to say that 'H'england was wery nice'.

17th November proved to be bitter-sweet for the Tristan community as it celebrated its first birth and its first death on the same day. 67-year-old Johnny Green died of pneumonia in Redhill Hospital, where in the maternity unit Johnny's niece Joan gave birth to her first daughter, Avril Hazel Elizabeth Repetto.

After someone dies on Tristan da Cunha all work comes to a halt, children and dogs are kept inside, and the entire community prepares for the church funeral and burial in the cemetery either the same or the next day. To maintain this Tristan custom, it was hastily arranged for Johnny's funeral to take place the following morning. Therefore, at 8.50am the next day four coaches carrying nearly 200 Tristan da Cunha refugees drew up outside the Church of St Mary the Virgin in the village of Bletchingley. The mourners were headed by their chaplain, Rev. Jewell, who was accompanied by a former Tristan padre Rev. Phillip Bell, with Peter Wheeler among the many expatriate officers attending. The Bishop of Southwark, Dr Mervyn Stockwood conducted the service, reflecting the prominence of the islanders' loss. After the funeral service and burial, everyone squeezed into the village hall for a 'stand-up' parish breakfast where there was simply not enough room for everyone to sit down and as was said jokingly at the time, even the sausage rolls had to be held upright. In that noisy crowd, however, the Tristanians and the people of Bletchingley began the process of getting to know one another, as the village families 'adopted' all the Tristan families for the day, inviting them into their own homes for lunch and tea. Later, as people returned to the church for evensong, Rev. Jewell gave an address in which he thanked the local community for the hand of friendship they had extended and pointed out that perhaps it was more than just a coincidence that the churches of Tristan and Bletchingley should both have been dedicated to St Mary the Virgin.

Peter Wheeler's first camp newsletter noted, possibly with some irony, that the entertainment available there rivalled that of Cape Town. He wrote that many of the youngsters sat for hours watching television, while Basil Lavarello and Bernard and Gerald Repetto had acquired second-hand bicycles and were frequently to be seen 'wobbling dangerously around the barrack square'.

Rev. Jack Jewell gave a talk about Tristan to Redhill Rotary Club. He emphasised how social equality amongst the islanders was exceptionally strong and expressed the hope that they would not lose here in England the gentleness and peace which they had nurtured in their community life on the island. In reply to questions, Jewell said it would be far better if the islanders could keep together as a village community in this country and that was their current hope. They wished to take their normal place in society here and he appealed for sympathetic consideration of their independent spirit. 'They are intensely loyal,' he said, 'and their most treasured possessions are photographs of the Royal Family.' Asked about his own experience as an SPG priest on the island, Father Jewell replied amid laughter, 'It was a revelation, for instance, for a parson to have a 100 per cent congregation. It was a very happy job indeed and I am very sorry it has come to such an abrupt end. I shall stay with them as long as I am needed.'

On 27th November Gilbert and Agnes Lavarello became the first islanders to move out of Pendell Camp, since Gilbert had secured a job tending cattle at the Royal Veterinary College. He earned £8 12s (£8.60) a week and was provided with a rent-free three-storey house at the desirable city address of 6 Royal College Street, London NW1. Agnes soon found a job cleaning at a local office, and when interviewed by the *Daily Express* she said, 'It's so different. We have never had stairs before, or noisy streets outside, but we are glad to have our own home again.' The house came with a snag, as Gilbert confided while piling coke on the fire, 'I couldn't get this stuff to light at first. I hadn't seen it before.' Gane's Tristan da Cunha Fund provided over £500 to furnish the house and cover the couple's removal costs.

A new life brings changed routines

On 4th December, the *Daily Telegraph* reported that, of the 164 Tristan men and women between 16 and 60, 117 were in work and already saving to buy possessions for their permanent homes in Britain. Several of them were interviewed. Edwin Glass was earning £9 a week loading lorries. He said, 'I miss farming, but I am glad to be doing this work. I have a little saved. I hope to add more to my savings for our future home.' Arthur Repetto was employed by the local council as a road sweeper. He said, 'I miss the sea. But it is good to earn money of my own. I am grateful to the people here.' Two of his three daughters had jobs in shops and one of his two sons was working as a labourer and saving quite a substantial amount. Whilst at Pendell the Tristans needed to spend very little because food and accommodation was free. Eight women and two men were employed in the camp canteen which was now run by an industrial catering company. The unit manager said, 'They are wonderful workers. They are willing to learn and if one of them falls sick, another member of the family turns up instead.' One thing was particularly significant: no one talked then of even the possibility of returning to Tristan.

Sadness again shrouded the community when 84-year-old Annie Swain died in an ambulance as she was being rushed to Redhill General Hospital on 9th December. Her son, Fred, was also admitted with the same bronchial illness. The following day Rev. Jewell baptised Avril Repetto in a simple ceremony attended by over eighty islanders in the old NAAFI headquarters at Pendell Camp. On a make-shift altar on a table covered with candles, a cut-glass bowl was used as a font. That same afternoon news came through to Jack Jewell that his own son was being born so he rushed to Redhill Hospital to greet the baby and his wife, only to be called away to attend at the bedside of 67-year-old grandmother Maria Lavarello who died of pneumonia that same afternoon. So, with the Union Jack flying at half-mast at the camp following a second death in two days the islanders again combined celebration with grief in the course of just a few hours. The media renewed its attention on the health of the evacuees. As well as three deaths, seven of them were in the camp's sick bay, an eighth in hospital with severe bronchitis, and many of the community were suffering from coughs and colds. In a statement Dr Samuels, their resident medical officer, said,

The Tristans were not exposed to this sort of illness on the island and are, consequently, unable to throw it off. They have received all the inoculations we have been able to give them to enable them to fight these complaints. The Medical Research Council will be conducting tests – not purely from an academic point of view – but to see how they can help.

On 12th December, the author Margaret Mackay made a visit to Pendell Camp and talked to the islanders. Mackay soon spotted scarfed and shawled women proudly scrubbing the front steps of their prefabricated wooden huts, a bearded older man in a wrinkled suit and sailor's cap, and a dark, handsome youth in a knitted hat buying sweets from the camp's small shop. Many of the islanders had never been on the Tristan mountain and so learning to walk around safely on the snow and ice was proving a new experience for them. One islander told her that he marvelled at the traffic and the towns but didn't enjoy walking along pavements amongst strange faces with 'nobody saying hello'. Taken by Rev. Jewell into the senior class under the charge of Rhoda Downer, Margaret found the children eager but homesick. They were excited at exploring their new environment, seeing their first horse over a fence during a nature walk, learning about the wooded countryside, and developing the ability to recognise oak, ash and elm trees. They were also able to identify the plentiful native birds including blue tits and blackbirds feeding at the bird-table set up outside their classroom. Mackay read essays written by the children after a day trip to London where they were shown the busy streets, introduced to the famous red double-decker buses, and taught various elements of road safety. Asked what most impressed them in the capital – Was it the glittering shops, Buckingham Palace, the guardsmen? – they again and again chose 'the River Thames'. When asked why, they replied 'because it was water, and there were boats', to which Miss Downer added 'they miss the sea'.

In the House of Commons on 18th December Cledwyn Hughes MP asked the Secretary of State for the Colonies if he would make a statement upon the present state of health of the former inhabitants of Tristan da Cunha now in this country. Reginald Maudling responded by saying:

Because of their isolation from viruses in their previous life the islanders are as yet peculiarly susceptible to colds and influenza. This vulnerability is in some measure accentuated by the prevalence of asthma among them. Three elderly islanders have died and at present nearly all the islanders have colds, and some are down with influenza and bronchitis, but there are no serious cases. Their health is being looked after by their own doctor who was working with them on Tristan and who came with them to Britain. He is working in co-operation with local doctors and the Medical Research Council who have arranged for detailed clinical investigations to be carried out under the guidance of eminent consultants.

Identifying the location of a permanent home.

There was no serious thought in December 1961, even amongst the islanders themselves, that the evacuees could ever go back to live on Tristan. All efforts were fixed on finding a permanent home for them in the UK. A report from the Colonial Office focussed on that challenge, the priority being to keep the Tristans together.

Many wanted to help and amongst the list of possibilities, these seven suggestions were interesting:

1. Stroma was suggested by Mr J Dale from Liverpool. This is an island lying in the Pentland Firth between the north coast of Scotland and the Orkneys. Formerly with 271 residents, the population had fallen to just twelve in 1961.
2. The stamp dealer John Lister was prepared to offer his own two 50-acre islands near Ullapool in north-west Scotland as long as local crofters could be provided with alternative grazing.
3. Mrs Dundas, a member of the notable property-owning family, suggested that 750 acres of the Dundas estate on the Isle of Skye would be a suitable place.
4. The Shetland Council of Social Service suggested resettlement in the Shetland Islands and agreed to instruct an officer to investigate their suitability, if desired.
5. Mr M Nicholson of Angus suggested that the Hebrides island of Rhum was the best option in Scotland, since it had empty houses available on it.
6. Sutherland County Council Planning Committee suggested the Durness Air Ministry camp near Cape Wrath on the coast of north Scotland.

7. The Bishop of Limerick through the Irish Government suggested resettlement in the Blasket Islands off the west coast of County Kerry. These had been uninhabited since the inhabitants were evacuated in 1953.

In assessing the rival claims the Colonial Office considered that all the northern Scottish locations had too bleak a climate, especially the Shetlands which were considered 'very poor.... The Tristan islanders would find it hard to make a living there.' Stroma was thought to be 'more suitable for a penal settlement', whilst Rhum was a National Nature Reserve and therefore not available. In the case of Skye both the cost of its purchase and the need to build houses on it rendered it too expensive. Nevertheless, the Colonial Office kept the options of Shetland and Durness under consideration whilst exploring suitable vacant service camps in England.

On 9th November questions were raised in the House of Commons from the MPs Richard Marsh and Cledwyn Hughes about what arrangements were being made for the resettlement of the Tristan da Cunha inhabitants. In response the Secretary of State for the Colonies, Reginald Maudling, explained, 'Various possibilities are being investigated and they will then have to be discussed with the people. It may well be some time before a final decision is taken.' The press was eager to know how the Tristan people were settling in, but, of course, Pendell was merely a transit camp that the islanders accepted while officials were pondering over the really big question: 'where next should they go?'

Since finding an estate of fifty to sixty unoccupied civilian houses was exceedingly unlikely, the search concentrated on identifying any empty service camp that, unlike the one at Pendell, would offer families proper decently separated housing.

On 17th November, in a significant move, Peter Wheeler was joined by former Administrator Pat Forsyth-Thompson and Gordon Whitefield from the Colonial Office to carry out their first visit to an RAF camp at Calshot in Hampshire where Southampton Water meets The Solent. The group planned to make a similar trip the following Monday to an army camp at Merebrook near Malvern in Worcestershire, both being promising locations that might be immediately available.

Following the two visits to Calshot and Merebrook, a meeting at the Colonial Office favoured the former. Merebrook was ruled out since, like Pendell, the accommodation there was in huts not suitable for families and it would require at least £800 for each unit to be upgraded. Merebrook was also regarded as relatively isolated with few employment opportunities within easy travelling distance. At Calshot 44 houses were already in existence, all built since the war to 'council house standards' with two or three bedrooms, while a further twelve with a single bedroom might be available. Close to the sea, the camp was somewhat isolated on a peninsula, but had an existing bus service, a doctor and a school within 4km. Good employment opportunities existed at the nearby Fawley oil refinery, as well as other local businesses and in Southampton which was 22km away. Several men went to view the houses in Calshot. They said that they would be happy to live there since they would be near the sea and able to stay all together as a community.

Despite the positive news about Calshot, plans were being made for Gordon Glass to fly at the expense of the *Daily Mail* to Kelso and on to the Shetland Islands, to report on possible resettlement there, and for Lindsay Repetto to fly to the Isle of Man which was now being considered as an alternative site. Allan Crawford wrote to the Colonial Office promoting the notion of the islanders settling in Scotland or on one of the Scottish islands, believing that they should be kept together as a group, that they would easily endure the colder conditions, and that there they could continue their productive activities around fishing, boat building and farming. He took the view, also generally held elsewhere, that the eruption had ruled out any thoughts of a return to Tristan and offered his services to help rehabilitate the islanders in the UK.

Gordon Glass returned from his trip to Scotland. Whilst there, maybe because he wanted to angle his remarks so as to sound suitably grateful, he gave the impression to the Lerwick correspondent of *The Times* that the Tristan islanders 'could very easily settle down [there] and become part of the community', but to his fellow-Tristans he reported that he was disappointed with both Orkney (where the land was heavily cultivated and no housing was available) and Shetland (where the south lacked accommodation and, though the north had housing, it was unattractive). Aunty Martha summed up Gordon's view when she later wrote that he had told her, 'It is not fit for us to live there. It is so cold and such bad weather for fishing, and so the people will never go over there.' Following Lindsay Repetto's visit to the Isle of Man, the Government in Douglas had taken an increasingly keen interest in the islanders and the matter of possible settlement had been put to the Manx Island Council. It was therefore decided at a Colonial Office meeting that, if Calshot should prove unavailable, the Isle of Man would be worth further consideration. If, however, the islanders wished to settle in Scotland, it was agreed that Wheeler and some of their number would pay a further visit to the area. On the other hand, should the move to Calshot be confirmed, Sir Irving Gane would launch an appeal the following week with a target of £30,000, that being a sum sufficient to provide £500 per house to equip it to the standard of an airman's married quarters.

Wheeler talked to the islanders and in a newsletter published in late November he confirmed the view that the Shetlands and the Isle of Man, each recently visited by Gordon Glass and Lindsay Repetto respectively, as well as other islands, were not practical for permanent settlement since there was neither work nor houses available in any of them.

On 19th December Colonial Secretary Reginald Maudling announced that the Tristan islanders would be moving to Calshot in January into fifty houses that had been offered by the Air Ministry. Now the islanders knew that they would be settling in Hampshire attention switched to making the most of Christmas in their temporary refugee camp as they prepared to make what was then considered a final move which at least would keep the community together.

Planning the Royal Society Expedition

By mid-November, a month had passed since the biologist Martin Holdgate had suggested that the Royal Society organise a visit to the abandoned islands. The Society – Britain's National Academy of Sciences – already had a 'Southern Zones Committee' chaired by Prof. Carl Pantin FRS of Cambridge University. Under it, Holdgate had organized and led a Darwin Centenary expedition to various parts of southern Chile that Darwin had visited in HMS *Beagle.* Following that venture the Society had held a major discussion meeting on the Southern Cold-Temperate Zone, with the biology of the Tristan islands as one of its topics, and it remained interested in the whole region. A Tristan da Cunha Committee was set up with geologist Prof. Lawrence Wager FRS as its chair, volcanologist Prof. William Q Kennedy FRS as its senior scientific director and Prof. Pantin as biological director. Holdgate was not involved as he was already committed to spending the southern summer in the Antarctic. With these senior scientific figures at its head plans were rapidly taking shape for an expedition to the Tristan islands in the early months of 1962 when good weather conditions were more likely to be experienced. The expedition would study the new volcano, make a wider geological study of the Tristan group, and investigate the effect of the volcanic activity on the vegetation and fauna. It was to be led by a geologist, Dr I G Gass of Leeds University, with three other geologists (one being Dr Roger LeMaitre, formerly of GISS). The Trustees of the World Wildlife Fund supported the project enthusiastically and made it possible for a botanist and zoologist to be included in the party. Allan Crawford, who was working full-time in the Cape Town Meteorological Office, agreed to be a consultant to the expedition, to give advice on shipping issues, and to feed-back reports received from any ships in the Tristan area. Later he offered to join the venture to reconnoitre a new site for a Tristan-based weather station.

Peter Wheeler announced the expedition plans in a newsletter to the islanders on 20th November, explaining that two geologists (LeMaitre and Dr P G Harris) would be making a reconnaissance visit to Tristan before Christmas aboard HMS *Jaguar*, and that the Royal Society planned to send a full-scale expedition in the New Year with Allan Crawford as a member and probably including Dennis Simpson and two islanders. On 6th December Roger Le Maitre (from the British Museum) and Peter Harris (from Leeds University Geology Department) flew to Freetown in Sierra Leone where they joined HMS *Jaguar* for the voyage to Tristan. Their remit was to examine the state of volcanic activity and gather information helpful to the main expedition which was scheduled to travel out in early 1962. LeMaitre's choice was significant because, as GISS geologist, he had spent six weeks on Tristan in 1955 and knew the island well.

In Cape Town on 11th December a special shareholders meeting of the Development Company was held in the aftermath of the volcanic eruption. Although the company had been in profit for the previous three years, this had not offset earlier losses. The company's recent successes had been attributed to the introduction of a freezing plant to replace the canning operation at Big Beach, but the factory had now been destroyed by lava. The firm held significant assets estimated at 260,000 Rand, including the two fishing vessels and a Tristan islands fishing concession with the UK Government that still had twenty years to run. However, the total collapse of its activities on Tristan was catastrophic for its continued existence and, rather than commit further capital to allow it to carry on trading, the meeting voted that it should go into voluntary liquidation. In the event a new company named Tristan Investments Ltd (Pty) was established under entirely private ownership. As a result, a new entity, operating as the South Atlantic Islands Development Corporation (SAIDC), was able to continue fishing activities in the region using the existing ships *Frances Repetto* and *Tristania*.

Early in the morning of Saturday 16th December HMS *Jaguar* arrived off Tristan bringing with it the Royal Society's advance reconnaissance team. Despite the gloomy dawn, light from the volcanic eruption could be seen from more than 20km away. Smoke issued continuously from the crater, and an immense cloud of steam rose from the ocean's edge where the hot lava still flowed into the sea. The original dome had split open and a new cone was growing on the seaward side, with two or three separate vents visible. From it a stream of lava ran seawards. Every few minutes thick yellowish smoke billowed from the volcano with an audible roar, and huge pieces of rock and molten lava were thrown nearly 60 metres into the air. The height of the volcano was measured and found to be 150 metres above sea level; it had doubled in size since HMS *Leopard* had left in October. The lava field extended nearly 400m beyond the original coastline and was about 950m wide at its seaward margin.

The village was completely intact and, so long as lava continued to flow eastwards and seawards, it would appear to be in no danger. The scene amongst the houses was orderly and peaceful. Cattle grazed quietly, chickens ran amongst them, but no dogs were seen. A tractor lay abandoned on the slope and, of course, there were no people. The air of normality was in stark contrast to the awe-inspiring spectacle only a few hundred metres away, where the factory lay under thousands of tonnes of black, jagged lava which smoked and steamed into the air. Big Beach was almost completely engulfed, and the main flow of lava had moved around Julia Point well into Little Beach from the east. Owing to the heavy swell no landing could be made, so HMS *Jaguar* circumnavigated the island looking for any other signs of volcanic activity, but none were to be found. On the ship's return to the north of the island a radar survey was made, and a detailed sketch of the volcano was completed, though the conditions still proved to be too rough for a landing.

By the following day, the wind had shifted but still blew strongly. The motor whaler was launched from the frigate and the two scientists with two of the ship's officers proceeded gingerly inshore to carry out a closer visual and photographic survey of the erupting volcano. The lava, which had a fairly solid and rocky crust, was estimated to be 20 metres thick at the water's edge. The lava had advanced appreciably since the previous day and at its edge the sea temperature measured 27°C compared to 15°C further out. As a landing was out of the question and time was short, the whaler was recovered, and HMS *Jaguar* set course for Simonstown. As the ship left, the Union Jack could be seen still flying over the Settlement. Lying on the beach were the islanders' remaining longboats, waiting for owners who, it was reasonable to assume, would never return.

On 21st December HMS *Jaguar* returned to Cape Town where Capt. Goodhugh reported that an estimated 30,000,000 cubic metres of lava had been thrown up, resulting in a major change to the coastline of the island. Dr Le Maitre predicted that, judging from other extinct volcanic cones on Tristan, the present eruption would probably subside, though this might take a few months, if not indeed more than a year or two. Crucially he added that 'once the volcano subsides it might be possible for the island to be resettled'. From someone who knew the island well and was a professional geologist, this remark of his was the very first indication of the possibility that the islanders might return home. It was not widely reported, and the subject of resettlement would be outside the remit of the expedition itself.

At a press conference back in Britain it was confirmed that preliminary arrangements had been made for the Royal Society Expedition the following spring. The venture would be led by Dr Ian Gass, an eminent geologist from Leeds University. Included in the team would be two island men, 31-year-old Joseph Glass and 25-year-old Adam Swain, both volunteers, who would act as guides. The choice of men was made by Chief Willie Repetto in liaison with Peter Wheeler. Both men were fit, had good knowledge of the sea and mountain and, as single men, were not responsible for families, so they were under no obligation to continue earning money to furnish any of the new homes in Hampshire.

Arrangements for the Royal Society Expedition were complete by 3rd January 1962. Dr Ian Gass would lead a team of twelve comprising four geologists – himself, Dr Peter Harris, Dr Roger Le Maitre, and Peter Baker, a zoologist (Donald Baird) and a botanist (James Dickson). These specialists would be supported by two radio operators, Staff Sgt Bob Shaw and Cpl Terry McCormack of the Royal Signals, as well as Adam Swain and Joseph Glass. The islanders also requested that two expatriates who they knew, trusted and had a detailed knowledge of Tristan, should join the team. These were Allan Crawford, who had been a member of the Norwegian Tristan Expedition in 1937/38. He had also served as meteorologist with the wartime naval station in 1942/43 and was already acting as consultant to the Royal Society Expedition; and Dennis Simpson, the agricultural officer who had been evacuated with the islanders and continued to work with them at Pendell Camp. Significantly, this was the first recognisable sign that the islanders would seek to assert their will, as by their lobbying that they ensured four team members were of their express choosing.

Since Crawford was based in South Africa, he played a pivotal role in liaising effectively with the authorities there, in securing equipment, and in acting as a link with the rest of the team as they gathered in the Cape later in the month. The Royal Society agreed to pay a surcharge on Crawford's life insurance policy, an indication of the perceived hazards that would accompany his work in the field. He was also asked to comply with certain press and publicity conditions which had already been agreed with the other eleven team members. When the expedition assembled in Cape Town, he was also appointed Deputy Leader.

The team planned that they would depart from Simonstown aboard the South African Navy frigate *Transvaal* on 22nd January and the Royal Navy had undertaken to send HMS *Protector* to pick up the party on 22nd March and return them to South Africa, thereby allowing them two full months ashore on Tristan to do their work. The expedition was ready to return to the active volcano.

The refugees face their first really cold Christmas

In advance of moving into new homes at Calshot the South Eastern Electricity Board arranged a series of lectures on cooking for the Tristan housewives, since it was expected they would be using electric cookers rather than paraffin or oil stoves with which they were more familiar. The women were also given talks on washing machines and electric irons, whilst the men prepared to attend a course on house-wiring and how to mend fuses.

Islanders were subjected to a series of medical examinations and tests. A mobile radiography unit from the local hospital authority conducted a routine check-up at the camp and islanders queued to take advantage of the opportunity. However, the Government was concerned that the islanders were perceived as guinea pigs for a whole range of medical research purposes, and this it wished to resist. It was therefore agreed that the first consideration must be the general health and well-being of the Tristans, and that any research should be limited. The Colonial Office planned to terminate Dr Samuels' contract, because the responsibility for the community's welfare would be transferred to the local health services when they had settled in Hampshire.

Christmas was approaching. *La Porte News*, the Fuller's Earth Copyhold Works company magazine, included a piece about Dennis Green and Johnnie Repetto. Although they had no previous experience of using the factory machinery, said the article, the two recruits had adapted well and were enjoying their work. Elsewhere, the Merstham Ladies' Club had raised sufficient funds to deliver gifts to the value of £30 to the camp for distribution on Christmas Day. These included a parcel of household linen for each of the sixty families, a silver bracelet for little Avril Repetto, and hot water bottles for the older people. Over that Christmas, island children were specially treated to two traditional entertainments up in London when they went by coach to see the spectacular Bertram Mills Circus at Olympia and the London Palladium pantomime which featured the comedian Charlie Drake as *Little Old King Cole*. Delights indeed.

The Tristan longboat *Carlene* was declared by the *Daily Mirror* to be the 'hit' of the *Daily Express* Boat Show held in London's Earls Court in early January 1962. The vessel, built of driftwood and canvas, was 'practically unsinkable and can nose into rocky places no other boat could attempt'. The paper reported that the islanders hoped to build smaller versions for British use and that three sailing clubs on England's south coast had expressed an interest in the idea.

Gordon Whitefield made the arrangements at Calshot for the islanders' forthcoming move from Pendell Camp on 23rd January. A total of 44 two- and three-bedroom houses were ready for permanent residence, twelve temporary quarters were available, and eight semi-detached one-bedroom bungalows would be ready for occupation within four months. The islanders would be accompanied to Calshot by a team of expatriates who would all leave after a short period of time, as once settled in Hampshire they would be treated as ordinary UK citizens with all the services to which they were entitled being provided by the local authorities. These expats were Peter Wheeler (who would continue in his role as Administrator until June and then return to Kenya after leave), Rev. Jewell (who would stay for a few months), the family of Dennis Simpson (who was by then a member of the Royal Society Expedition though he would be returning to Calshot in March), and Dr Samuels (who was given a month's notice the previous December). In fact, Samuels had given up his Cape Town home to work on Tristan and was unemployed and homeless at the end of January 1962.

The Calshot houses were being transferred from the books of the Air Ministry to the Ministry of Works for a nominal value of £52,000 which was to be paid by the Treasury. Later they were to be taken over by the New Forest Rural District Council as part of their rented housing stock. The Treasury agreed to the expenditure of £14,450 to provide basic furniture and electric cookers, and the local WVS were organising a delivery of basic foodstuffs to get each family started in their new home. Gane's Tristan da Cunha Fund provided money for the WVS to buy this supply of food, as well as to purchase curtains and enough coke and coal per family to last for the rest of the winter. The WVS had also asked local people to donate good-quality surplus linen and similar items to be available for the Tristanians on arrival. They encouraged residents in the area to introduce themselves to island families and to see whether there was any help they could give as they settled in.

In Bletchingley Church on Sunday 14th January Jack Jewell gave a farewell sermon and showed a film of the erupting Tristan volcano made the previous month by the Royal Society Expedition advance party. The local congregation were fascinated to see the volcanic activity taking place almost at beach level rather than on the mountain top crater. The rector, Rev. Ron Brownrigg noted, 'I think this film will have convinced the Tristans of the wisdom of remaining in the UK for some years to come.'

There was a flurry of social activity in Surrey before the islanders' departure. This included three islander weddings in Bletchingley Church where on 16th January Peter Repetto married Pat Swain and four days later the double wedding when Benny Green married Sylvia Repetto and Lars Repetto married Trina Glass. On both occasions crowds of locals clustered around the gate to view the newly-weds and receptions were held at Pendell Camp. Rev. Brownrigg commented that perhaps one of the most memorable links with the Tristans had been the sequence of weddings in the final few days as the church was crammed with islanders and locals side by side. Unruffled by the TV and press cameras the bridal parties maintained the traditional poise and simple dignity that the local community had come to recognise and respect. As on Tristan, there was no departure for a honeymoon, but proper new homes awaited in Hampshire in a few days' time.

Photograph of the reception held at Pendell Camp on 20th January 1962 to celebrate the double wedding of Benny Green and Sylvia Repetto (right) and Lars Repetto and Trina Glass (left). Lars and Benny were later chosen to be members of the Resettlement Survey Party, leaving their brides at Calshot whilst they made a start on building new homes on the island for their families. (Trina Repetto)

The time for moving was now getting close. Peter Wheeler gave what was described in the local parish magazine as a wonderful party to the many groups and individuals who had helped the Tristan community while at Pendell. Many parishioners attended and, in a reversal of roles, they were entertained by the Tristanians. On behalf of the people of Tristan, a model longboat was given 'with gratitude' to the parishioners.

The Royal Society Expedition sets forth

The two expedition radio operators left England on 5th January to begin an epic air journey to Cape Town which included stays in France, Malta, El Alamein, Khartoum, Aden (in what is now Yemen) and Nairobi covering twelve days. On the same day MS *Crestbank* passed Tristan and reported that the volcano was actively spouting smoke. Steam rose as the molten lava, now stretching 1200 metres along the coastline, continued to invade the sea, though the Settlement remained untouched.

Allan Crawford, assisted by Norman Vaughan (working for the travel agents Thomas Cook in partnership with the Crown Agents), was busy in Cape Town buying tonnes of food, camping kit and other stores for the forthcoming expedition. A key purchase was of a 3-metre fishing dinghy which Crawford arranged to have painted fiery red and appropriately named it *Vulcan*. Such a good lightweight boat would be essential in survey work around the island, now that it was devoid of the two main sandy landing beaches near the Settlement. The stores were packed in wooden cases, each weighing no more than about 25 kilos. This constraint enabled them to be man-handled on the steep rocky shoreline which had now replaced Big and Little Beaches on the island.

Interviewed at Cape Town airport on Friday 19th January, Adam Swain said that the islanders were getting on well in Britain but still yearned for their old home. He and Joe Glass hoped that the expedition might give some clues as to whether they would be able to return. The following day the complete Royal Society Expedition team met for the first time at a Cape Town hotel and on Sunday 21st January a reception attended by the press and local dignitaries was held at the British Minister's house in the city.

At 2pm on Monday 22nd January, the 1460-tonne frigate SAS *Transvaal* departed Cape Town and the expedition was properly underway. The small vessel was not designed to accommodate extra passengers, and so space was cramped. However, the voyage got off to a successful start in calm conditions with good weather which lasted throughout the following day before strong south-westerly winds confined many of the men to their cabins. Other passengers aboard were an army doctor, media representatives and two police detectives, sent by the Ministry of Foreign Affairs in Cape Town to investigate reports of looting on the island, to search any South African fishing vessels operating in Tristan waters, and to warn their crews not to go ashore.

January 1962: settling in at Calshot

The islanders' move from Surrey on Tuesday 23rd January to the former seaplane base in Hampshire was understood to be the final stage of their enforced evacuation which had started fifteen weeks earlier. They were now accustomed to England and about to start a new life near the sea in decent homes with good prospects of employment and a full range of services in an affluent part of the country.

It was a total of 262 islanders who now moved into 64 homes in a road that had been re-named Tristan Close. Family groups ranged from just two in a one-bedroom bungalow to no fewer than eleven in the case of Cyril and Ellen Rogers' family who squeezed into 1 Tristan Close. A priority for all the families was staying together, even though that meant overcrowding until further housing became available. Others would follow in due course. Adam Swain and Joe Glass were currently aboard the *Transvaal* in the South Atlantic and would re-join their families later in the year. Six islanders were still living in the Falkland Islands, although five of them would later move to Calshot. Gilbert and Agnes Lavarello remained in London.

The move went smoothly and within an hour the islanders were sitting down to a meal of brown stew and fresh fruit prepared by the WVS New Forest East Emergency Feeding Team. The islanders themselves had chosen the menu. Interviewed by *The Times*, Peter Wheeler confirmed that the support team's main objective was to 'ease out as painlessly as possible. We are relying on local voluntary organisations to make it easier for the islanders.'

Eighteen senior pupils started their studies at Hardley Secondary Modern School in nearby Holbury. The Tristan da Cunha Fund provided half the cost of their school uniforms, and Rhoda Downer, their previous teacher on Tristan and at Pendell, joined the school's staff. In contrast, the 22 younger pupils who travelled to Blackfield Primary School were about to begin their first day without the reassurance of their familiar teachers. The Hythe Branch of the Ministry of Labour moved a team into an office at what was still being referred to as 'Calshot Camp' to help Tristan adults find employment. It is hard to grasp the sense of novelty felt by these families as they settled into their new houses, climbing stairs to bed for the first time, and adapting to lino floors, fridges, and electric cookers. In contrast to their Surrey accommodation where all the basics had been provided, the islanders now had to face 1960s UK economics as they attempted to manage their own budgets and cope with paying their rent and rates of between 30s (£1.50) and £2 a week. At least they could draw some sense of reassurance from the fact that a former RAF chapel, a community hall, Post Office and store were within easy walking distance.

Now that the islanders were living in an ordinary residential area any controls on the TV or press were impossible. Soon they became not only increasingly wary of what they saw as an invasion of their privacy but had started to grow antagonistic towards it. One of the many media correspondents was *Daily Mail* reporter Vincent Mulchrone who made an early visit to Tristan Close and in his words 'saw the Tristans become British' as they were given 'dreary pebble-dash "semis", modern furniture, electric cookers, National Assistance, rates and income tax – all the blessings they've had to do without for 147 years'. He thought they were being swamped beneath a sea of green WVS uniforms, suffocated by sympathy and choked by kindness, to which they were reacting with their typical wide-eyed humility. He continued, 'Seldom can refugees have had better treatment,' as each house was fitted with £300 worth of furniture including brand-new beds and linen, as well as sparkling pots and pans. Yet, in his view, the islanders were not truly contented, since the original notion of transplanting them onto another island where they could farm and fish had been quietly dropped, and their traditional way of life was therefore now over. Instead, the Colonial Office had decided to end the Tristan story by integrating them into British society, with communal activities no longer encouraged and children going to local schools. This, according to Tristan elders, would inevitably lead to the community eventually splitting up. In summary Mulchrone thought the Government could not have done more, except perhaps that the use of a little extra imagination might have helped.

Of course, the dream island where the Tristans could replicate their communal farming and prosperous lobster fishing didn't exist and a move to Scotland was ruled out by the islanders themselves rather than by the UK Government. By moving to Calshot in south-east England, the community would stay together and remain in the media spotlight long enough to be in a much stronger position, when the time was right, to push for a return to Tristan da Cunha.

On Wednesday 31st January Sir Irving Gane travelled to Calshot for a meeting with Peter Wheeler and the islanders to review the assistance provided by the Tristan da Cunha Fund and to consider what further help it could offer. It was agreed that it would supply carpets for all the living rooms and stairs, and rugs, quilts, and bedspreads for the bedrooms. Samples were available for families to choose their own fabric and colours. With an aim to make the islanders more independent, it was also agreed to provide carpentry and gardening tools, and a garden shed for every home. A 1.2-hectare allotment area was later acquired for their use, with the fund covering the rental charge, the initial cost of ploughing the land, and a stock of seed potatoes. Andrew Glass became the second islander to secure a job, in his case attached to the Royal Veterinary College working with farm animals at its rural Hertfordshire campus. The Tristan da Cunha Fund came forward with £149 for the removal expenses for Andrew, Ivy and their son Eric to move to their new home at nearby Brookmans Park.

Friendly neighbours soon got to know the people who they called the Tristans as they became familiar faces in local shops, churches, and pubs in Hampshire. When local people heard of plans for some Jehovah's Witnesses from Southampton to call on Calshot homes, they consulted the Anglican and Roman Catholic priests in the area, and it was decided that their Tristan congregation should be warned about the expected visits. Apparently, the islanders were not particularly accommodating when those smartly dressed evangelists arrived with briefcases in hand, an untypical reaction from a people widely renowned for their hospitality and friendliness.

England's winter in 1961/62 was exceedingly cold, and the islanders were soon having to deal with burst pipes. A local plumber from nearby Holbury was obliged to visit many of the houses to deal with the leaks and to instruct the new residents in how to use the local telephone box to call him in an emergency. Also, on hand to help were John Tucker, the Senior Welfare Officer with the New Forest Rural District Council, and his colleague Mrs Calvert, both of whom befriended the islanders. In addition, every week a representative from the WVS visited each house to help deal with any problems and provide advice on form filling, a task that was ever present but novel to the islanders.

Nor were they immune to upsetting events. Gordon Glass, who had paid that visit to reconnoitre the Shetland Isles, had lost an arm as a result of a machine-belt accident at the fish factory on Tristan. Now 62, he was limited in his options for local work, but had taken on the job of night watchman, keeping an eye on equipment at a local road works. Sitting as usual outside his hut, two aggressive youths, referred to as Teddy Boys, sneaked up on him, beat him up and searched him. Finding only an old family watch, they threw it in the road and stamped on it, before running off into the night. The incident, early in their time at Calshot, made the islanders fearful of going out in the evenings. Tensions were raised higher when groups of Teddy Boys began to arrive in Calshot on Saturday nights aiming to disrupt parties arranged by Tristan youngsters. Another illustration of the islanders' lack of worldliness occurred when a large bag of potatoes was bought in a Southampton shop and left outside its main door to be collected later when other purchases had been completed. Upon returning the islander was surprised that the potatoes had disappeared, and at first simply assumed that someone had taken them back into the shop. Lessons were quickly learned through experience, but the implicit trust enjoyed by all in their village on Tristan was sorely missed.

The group assembled outside Buckingham Palace on 9th March 1962 before meeting HM the Queen and presenting her with the island longboat *Carlisle*. Chief Willie Repetto is far left with his sister Martha Rogers, Head of the Women, alongside. Rev Jack Jewell stands in between the other women at the front, while Administrator Peter Wheeler may be seen far right. (Peter Wheeler)

Gordon Whitefield of the Colonial Office was still based at the headquarters of the New Forest Rural District Council in Lyndhurst. He reported that, despite some problems, 'the morale of the islanders is higher than at any time since their arrival in this country'. Many of them had expressed to him their gratitude to the British Government for providing such agreeable houses and furniture and had spoken of 'their pleasure at being able to live so close to the sea again'. Following the termination of Dr Samuels' contract, the islanders were now attending a local doctor's surgery, and Gordon Whitefield was able to report a great improvement in their health since their time in Surrey. Initially help had been provided in the use of the local bus service, but this was phased out as the Tristanians steadily gained confidence, especially on regular daily commutes to work.

The islanders settled into a wide range of jobs: working in garages and service stations; on various factory assembly lines, including 21 women in a Lymington food-processing firm; in road construction and street cleaning; as seamen on Union Castle ships; in a yacht club near Lymington; and at a boat-builder's yard at Hythe. Some islanders were employed by the Motor Museum and Beaulieu Manor sawmills, both owned by Lord Montagu. Eight women worked in the museum café, while Ken and Tim Green kept the vintage cars gleaming and acted as guides. Ken Rogers was employed at Thick Butchers in Fawley, becoming in due course a skilled and much-respected employee. When he was later back on Tristan, he would be sought out for his expertise in butchering meat into choice cuts, a skill he had learned while a refugee in England that would go on to enrich the future community on the island.

Perhaps the highpoint of the Tristan islanders' time in England was on 9th March 1962, when a deputation of islanders led by Chief Willie Repetto and accompanied by Peter Wheeler and Rev. Jack Jewell visited Buckingham Palace to meet the Queen and present her with the Tristan longboat *Carlisle* which had been brought from the island on the *Stirling Castle*. Shortly afterwards the boat was transferred to the National Maritime Museum in Greenwich and later to their collection in Falmouth where it is stored and sometimes displayed.

Part 5:
Making Resettlement a Reality
1962-1963

The Royal Society Expedition's work on Tristan da Cunha

At 9am on Saturday 27th January 1962 the Royal Society Expedition sighted Tristan da Cunha and approached to view the volcanic eruption. Activity had diminished since the initial inspection on 16th December, but white smoke issued from the main cone, enlivened by occasional explosions from within which sent up large jets of sulphurous vapour. Since the weather was too bad to disembark, the Transvaal sought to find the lee at Stony Beach and there try to make a landing.

In the event, this was not possible that day, but at 7am on 28th January Dennis Simpson, Bob Shaw and Adam Swain were taken ashore at Stony Beach using the dinghy *Vulcan* for the first time. Stores were left at a camping hut and the three set off along Deadman's and Seal Bay Beaches to the Caves on their way to the village. When climbing Gipsy's Ridge above Hackle Hill the weather closed in and the heavy rain forced them to abandon their trek and return to Stony Beach, picking up some potatoes stored in one of the Caves camping huts on the way. The weather was too rough for the trio to return to the *Transvaal*, which again anchored off Stony Beach that night. The following day the *Vulcan* again was unable to be used in the swelly sea conditions, as the wind had now shifted to the south-east, so the *Transvaal* returned to the Settlement, effectively marooning the advance party of three who hoped to be already at the village.

Back at the Settlement anchorage on 29th January, Ian Gass, Roger Le Maitre and Joe Glass boarded *Vulcan* and they gingerly edged their way towards the west end of the lava flow. At Garden Gate there was direct sea access onto a smooth wave-cut platform formed of rock pools which offered at least a modest amount of shelter. For the rest of the day crew and expedition members ferried the bulk of the expedition's stores ashore using an inflatable rubber pontoon which proved ideal for the purpose. The expedition had landed.

The emission of lava from the volcano appeared to have ceased, though sulphurous vapour still rose from the main cone and a subsidiary crater lay behind it. A final burst of vapour and dust was emitted symbolically on this first day ashore. During the short

Offloading stores from *Transvaal*, anchored behind, at Garden Gate Beach made to look easy on this calm day. This area would soon be covered with a boulder beach, which meant that the expedition had a falsely favourable impression of a good new landing place now that both Big Beach and Little Beach were covered by lava. (Allan Crawford)

The dinghy *Vulcan* being lifted ashore at Garden Gate by Allan Crawford, Jim Dickson, Joe Glass and Roger LeMaitre. (Donald Baird)

December visit such incidents had occurred at approximately ten-minute intervals and had again been noticed on the expedition's arrival two days before. In the dark that night a red glow could be seen coming from the central cone and from cracks in the congealed lava mass flowing from it. More glowing patches of lava were revealed as chunks of hardened rock broke away. Allan Crawford described the eerie atmosphere in the deserted village as he approached the Administrator's quarters to prepare a base for the expedition. Looking into the houses he was dismayed to see scenes of abandonment and desolation. Front doors were open, cupboards and chests of drawers had their contents scattered over the floor, whilst the locks of boxes and chests had been forced apart. Dead cats and chickens lay about, windowsills were black with dead flies, and papers and books were strewn in every direction. Given the methodical clear-up documented by HMS *Leopard*, it was clear that intruders had entered the village to steal belongings. The station was cleared room by room, a big bonfire was lit to dispose of what by then had become rubbish, and three bedrooms and a lounge were made ready as headquarters for the expedition in the Residency.

It was not until Tuesday 30th January that the 'advance' trio marooned at Stony Beach could join their colleagues. With conditions still too rough to board the *Transvaal*, the men set off each carrying heavy loads weighing some 15kg. They covered the beach walk to the Caves and climbed to the Base in two hours before setting off northwards to Burntwood. There they 'sledged' down the sandy scree by First Gulch following the locals' habit of using boughs of island tree to act as a toboggan. As they walked across the Patches Plain, they were met by Joe Glass, who, with Adam Swain caught four donkeys. So now they rode in some style up the valley behind Hillpiece and back to the village. Along the way they found a sheep's carcass with a bullet in it and were struck by how quiet the Patches Plain had become without the familiar bleating of the flocks. Later it was discovered that only seventeen of the 740 sheep abandoned just four months earlier had survived the Tristan summer. The hapless beasts had either been taken aboard marauding vessels for food or eaten by the dogs remaining on the island. Back in the village the navy team erected a derrick with a block-and-tackle device at the edge of the cliff above Garden Gate Beach to haul the stores up the 45-metre rock face. Whilst that was happening, Allan Crawford took one of the police detectives around the houses to show him the evidence of looting that he had discovered.

On their arrival in the village, Shaw and McCormack started to clear up the radio shack and managed to get one of the transmitters working to make a first contact with the Gough Meteorological Station using Morse code. Dennis Simpson set to work repairing the island's water supply. The Big Watron rises in a strong spring south-east of the village, just west of the new volcanic centre. Earth tremors and rock falls had dislodged the pipes which normally siphoned off the flow into a holding tank and supplied the Settlement with water of the purest quality. These were speedily re-connected, and the team soon had that vital resource running again.

By the following day, the expedition team had set up their headquarters in the Residency. Eight of their number would be able to sleep there, whilst the other four were housed in the adjacent station hospital. The hospital group consisted of Allan Crawford

Members of the Royal Society Expedition gathered outside their HQ at the Residency showing from left to right -
back row: Terry McCormack, Dennis Simpson, Peter Harris, Peter Baker, Bob Shaw;
centre row: Roger Le Maitre, Jim Dickson, Ian Gass, Allan Crawford, Donald Baird;
and front row: Adam Swain, Joseph Glass (Geoffrey Dominy)

and the Stony Beach gang of Simpson, Swain and Shaw. Foam-rubber mattresses had been brought and these were placed on the beds for added comfort, but unfortunately there was an infestation of fleas. This was blamed on the yet-unidentified looters who had ransacked the Settlement and, by leaving many of the doors open, had allowed farm animals access into buildings.

The island still had its tractor, that ancient machine left behind with the departure of the Royal Navy at the end of the war and sometimes known as 'Old faithful'. Dennis Simpson worked in the afternoon to get it going again so that it could be used to haul a trailer and ease the transfer of supplies from the cliff top to the team's base at the Residency. Soon, after the tractor was started, the diesel-powered island generator was working once again and able to provide electricity to power the lighting system and radio equipment. As a result, a first scheduled daily radio contact was made with the Royal Navy Radio Station at Youngstown near Cape Town.

At 9.30 on 1st February, SAS *Transvaal* departed for Simonstown, giving a dramatic send-off in the form of blasts from its horn accompanied by a pyrotechnic display of redundant rockets. As the frigate steamed by it dipped its flag, with the shore team returning the compliments by lowering the Union Jack now flying proudly once again outside the Prince Philip Hall. The two police detectives had come to no conclusions over the looting and had made no contact with the *Tristania* or *Frances Repetto* who were fishing the outer islands during their visit. Now the Royal Society team were able to fully focus on their scientific tasks. Whilst the geologists planned their mapping survey, Crawford, Simpson and the two islanders brought a dinghy from the beach and took it to the Bluff by tractor and trailer. While out west they attempted to stalk about eight dogs that they had spotted, but their .22 rifle was inadequate, and this awful task would have to wait. The following day the four geologists were busy studying the still-erupting volcano, burning their boot soles as they moved around. Using a pyrometer near the crater Peter Baker measured temperatures of over 850°C whilst recording an air temperature of up to 200°C above. Allan Crawford sent his first meteorological reports to South Africa, thereby renewing the long-established service of transmitting regular weather recordings to the Cape.

The excellent weather the team had enjoyed for a few days broke on 3rd February and the prevailing north-west winds resumed with a low cloud base and steady rain. Crawford drove the tractor to the Patches, taking with him the four geologists plus emergency food supplies and a tent to be erected there. Shaw and McCormack set up a wind generator at the radio shack to charge the radio batteries which were running low and could not be rendered workable using the main current. A 24-volt / 10-amp

Donald Baird stands in the doorway of the house which had belonged to Dennis Green, the only one on the island damaged by the volcanic eruption. The main walls were intact, but it was assumed a lava bomb had set fire to the flax roofing and destroyed the timber fittings, whilst the outbuildings beyond and the sentry-like toilet in front of the left side of the house were intact. (Sir Martin Holdgate)

View taken from the cliffs looking down into the active crater of the 1961 volcanic cone with the new inlet at Garden Gate Beach behind. (Donald Baird)

current was created when the wind gusted. Dennis Simpson and Joe Glass killed three more dogs but reported seeing a pack of about a dozen marauding around.

During their time on the island the team 'lived like fighting cocks', to use the words of Allan Crawford. Meals were excellent since a good supply of tinned goods and groceries had been left in the abandoned canteen. Although the Potato Patches, where cultivation had started earlier in the season, had become overgrown with weeds, potatoes could be dug from some gardens in the village. Terry McCormack caught and cooped about fifteen hens which Dennis Simpson attempted to cajole into laying eggs. He and Joe Glass killed a young calf for fresh meat, so the team enjoyed some splendidly succulent steaks for their first Sunday supper ashore.

The weather improved on 8th February, so a party of Dr Peter Harris, Adam Swain, Peter Baker, and James Dickson headed for the Peak. They aimed to camp on the Base for three nights and were transported by tractor and trailer driven by Terry McCormack to the Bluff before climbing Burntwood. A second party comprising Dr Ian Gass, Joe Glass, Roger Le Maitre, and Donald Baird departed aboard the *Vulcan* for a few days' study based at Sandy Point. In the event, the Peak party returned next day as it was so wet on the mountain.

The volcano was becoming more active again with swelling on the eastern ridge, plenty of rock falls and an increase of smoke emitting from the crater. During the evening of 10th February red cinders showered down the side as a block of lava broke off, and after dark it glowed more distinctly than at any time during the expedition's stay on the island. The situation was undeniably fearful, and with the threat of a further eruption, food and tents were taken to the Potato Patches as a contingency. The remaining dogs continued to cause anxiety since they posed a constant risk to calves and sheep. It was estimated that nine or so of these strays were left, growing increasingly feral as the cattle became better protected in the Settlement area.

On 15th February Dr Ian Gass, Adam Swain, Jim Dickson, and Bob Shaw went by boat to East Jews Beach to pick up rock samples previously left there. An elephant seal was hauled up on Snell's Beach, which Gass moved off to be able to walk along the coastline to Rookery Point, while he and Dickson collected more plants and rocks. Saturday 17th February was Adam Swain's 26th birthday so Joe Glass and Roger Le Maitre went fishing and caught seventeen lobster which were later served at a celebratory supper along with home-made bread and fried mushrooms freshly picked that day. A few days later Dennis Simpson and Joe Glass killed a bull weighing 90 kilos which was refrigerated and provided beef for several weeks.

Sunday 25th February was the hottest day so far and the fishing vessel *Frances Repetto* arrived with its captain, Geoff Dominy, who came ashore to pick up 1000 fishing nets and to gather together any mail waiting to go off to South Africa. Two days later the *Tristania* appeared at 8.30am and a party of eight went aboard for a survey of Nightingale and Inaccessible Islands. Only Allan Crawford, Bob Shaw, Peter Baker and Dennis Simpson remained on Tristan. McCormack's attention to his chickens was beginning to pay off, and by the time the others departed they produced a record ten eggs in a day. With only four to be fed, those left behind were now assured of a good breakfast each morning.

On 1st March Shaw and Crawford went up the path west of Hottentot Gulch to the 300-metre mark where they selected a big, rounded bluestone boulder which they painted white to act as a beacon and assist in estimating cloud height during weather observations. This landmark is still known as the White Stone and is a familiar sight above the village. Next day the *Tristania* reported that the Nightingale Island team had landed safely but it was too rough to land on Inaccessible. Gale-force winds drove huge waves ashore and it was calculated that about 3 metres of the new lava front was eroded in a single day, in the process sending clouds of steam up into the air as hot lava was exposed by the waves.

The 1961 lava flow from the Base with the village beyond. In the centre foreground is a newly formed lagoon which would be short-lived, and to the right the eastern edge of what was Big Beach. The central embayment in the lava flow provided a temporary sheltered landing place. (Donald Baird)

Adam Swain and Joe Glass at Stony Beach.
Note the ingenious way in which Joe Glass has tucked a stock of potatoes into his shirt.
(Donald Baird)

Intense activity continued over the next few days. On 4th March Bob Shaw took Dennis Simpson and Peter Baker by tractor and trailer to the Bluff before they climbed Burntwood for further study on the Base. Two dogs were seen by the Potato Patches and Shaw put out poisoned meat to attempt to kill what he thought, mistakenly, were the last of the animals. The following day, Dennis Simpson came off the Base for some rat poison since the wretched rodents were eating all the tents away. He also needed another sleeping bag as it was so cold on the Base at night. Also, on this day, the *Tristania* returned with the offshore party who had still not managed an Inaccessible landing.

The volcano had now become noticeably quiet; it was smoking less, no longer glowing at night, and there were scarcely any more rock falls. So, there was a genuine hope that the lava may well have stopped flowing. On 8th March Peter Baker ventured right up the east side of the new volcano from the Pigbite direction. This was obviously considered high-risk as Allan Crawford noted that he 'lived to tell the tale'. After lunch others joined him on the same excursion to the top of the new volcanic lava dome. Here it was still hot under-foot and immobile red-hot rock was visible in a fracture on the summit about 2.5 metres below the surface. There the temperature was determined at 890°C and the air temperature immediately above the dome at 35°C.

By this time the botanist, James Dickson, was well ahead with a study of the effects of the eruption on Tristan flora. He noted that landslides and rock falls on the main cliffs behind the Settlement and in Hottentot Gulch had removed and damaged vegetation. They had also reduced soil binding and made the area vulnerable to future mass movement. Dickson calculated that about 8 hectares of rough grazing land had been destroyed by the lava which had also obliterated Julia Point, an area known to be rich in marine algae. Volcanic ash covered about 4 hectares of the lower slopes of the main cliffs up to 30cm in depth. Pyroclastics emitted from the volcano, including lava bombs, had caused fires, especially among patches of the island tree up to 300 metres above sea level on the main cliffs and to New Zealand flax in the Settlement. Toxic fumes had the biggest impact on vegetation and provided the most unusual findings. First, the area of fume damage was determined by the prevailing north-west winds, which meant that upwind towards the Settlement, damage extended only about 1.5 kilometres, but this included all the village which therefore would have been uninhabitable during the main eruptive phase. In contrast, downwind damage extended to Sandy Point, 10 kilometres away. This whole area of toxic fume damage, which had mainly occurred between October 1961 and January 1962, covered the Plain, main cliffs and just the lower levels of the Base. The details of Dickson's subsequent report made it clear that the population was very lucky to escape the eruption unharmed, the decision to evacuate had been completely sound, and that the Royal Society Expedition team themselves were very fortunate in the timing of their survey.

Examination of the fumarolic minerals confirmed that sulphates, fluorides, chlorides, carbonates and borates were the major components of the toxic fumes released by the volcano, and that this admixture was dominated by the gas sulphur dioxide with lesser amounts of hydrogen sulphide, sulphuric acid, hydrogen fluoride, hydrochloric acid and carbon dioxide - a highly toxic mixture. As the islanders left by longboat on 10th October 1961 it was a brew of these gases which was within what appeared to be 'smoke' seen from the summit of the fledgling volcano and cascading down the north-west flank towards the village. If they had been enveloped within that apparent 'smoke cloud', people would have suffocated.

On 11th March five of the team went up on the top of the new volcanic cone and climbed down into the three craters where they found the conditions were much hotter than on top of the lava dome and the smell of sulphur fumes affected their breathing. Nevertheless, three days later, Bob Shaw was able to strike a positive note in his diary where he wrote, 'The volcano is very quiet now; no one pays much notice to it anymore.'

By now, the zoologist Donald Baird's work on the volcano's effect on the island's fauna was almost complete. He noted that, as the eruption occurred on the Settlement Plain, the vegetation was greatly modified. Bird life had sustained little damage since it was almost non-existent, the area being dominated instead by predators (cats, dogs and rats), grazing domestic mammals, and alien invertebrates. The new lava appeared to have destroyed the endemic insect *Tridactylus subantarcticus willemse* which had been collected only in that area. Immediately east of the lava field, where ash cover was general, the soil and ground-surface fauna were completely destroyed. The first living non-flying invertebrate, an alien centipede, was found 200 metres from the crater, under a stone buried by 2cm of ash. The first live worms were found 250 metres from the new volcano in moist soil overlain with 15cm of ash. West of the lava in the Settlement area, ash cover was negligible with little effect on animal life. On the Base, despite vegetation being damaged, invertebrates were little affected. Baird concluded that the eruption had a very slight influence on the alien and native fauna of the island, just as had been the case with similar smaller eruptions in the past.

The expedition team wind up their efforts

By 15th March, the endeavours of the expedition team were drawing to a close. The *Tristania* returned that morning and soon afterwards the Dutch liner *Tjisadane*, which had so crucially picked up the islanders from Nightingale on 11th October the previous year, was sighted as it steamed by. Dennis Simpson and Adam Swain returned to the Base to bring down the remainder of the stores left up there, salvaging what they could, though the tents had been damaged by the wind, and rats had eaten into the stores. All that was left for the team to do was to arrange an evening party on the 17th to mark the end of all their fieldwork.

The expedition's timing had been perfect. Safe field work within easy walking distance of their base had been possible at the very point when the eruption had slowed down and come almost to a halt. Observations, made at a time when the volcano and lava flows had virtually attained their maximum, provided a definitive record of the situation. Fumarolic activity would continue for many years, but the main predicted future change to the new volcano was erosion by the sea that would quickly shrink the lava front. The Royal Society survey identified three separate lava flows emanating from the central crater. The first and smallest ran straight down the slope and obliterated the factory at the west end of what was Big Beach. The second and largest flow followed the same course, but then diverged to swing across Big Beach towards Pigbite in a series of lobes. The third and final western flow occurred when the eastern flow had created a ridge, so lava found the lowest course and moved north-west from the end of the central dome. It had then covered Julia Point and run down to smother Little Beach from where the islanders had launched their longboats. At the far west of Little Beach, a wave-cut platform of smooth older lava remained uncovered by molten rock thereby providing for the expedition a convenient landing site at Garden Gate Beach.

Early on 20th March HMS *Protector* arrived and sent ashore a helicopter which landed at about 7.30am to begin loading some 500 kg of stores, samples and specimens and lifted off the personnel, an operation that was completed in two hours. Aboard the *Protector* was Martin Holdgate, who had proposed the idea of a Royal Society Expedition the previous October, and who now eagerly grasped the opportunity to pay a brief visit to climb to the top of the new volcano with Roger Le Maitre. The two of them also took aerial photographs of the cone and lava from *Protector*'s helicopter. In a final gesture a tame cat named Tristan was taken off the island to be given a home by Allan Crawford back in Cape Town. Since no landing on Inaccessible had been made by the team, the *Protector* sailed over to the island. The following day its helicopter carried a party of four to Blenden Hall for a visit of just over two hours ashore by way of compensation for their earlier failure. In the event, any meaningful fieldwork was impossible, especially since the organisers of the trip had failed to include either of the islanders who knew the rough terrain, dominated by stands of tussock grass absent on Tristan and a mire later named Skua Bog. However, the helicopter eased the group's disappointment by later circling over both Inaccessible and Nightingale Islands, so that a series of aerial photographs could be taken before departing.

Aerial view from the helicopter of HMS *Protector* on 20th March 1962 showing the 1961 volcanic cone and lava flows at their maximum. (Roger Le Maitre)

On 26th March HMS *Protector* arrived at the Simonstown Dockyard and soon afterwards the Royal Society issued a press release written by its expedition leader Ian Gass. Significant at the present stage of this chronicle are the following two sentences:

Categorically excluded from our terms of reference was the social problem concerning the possible return of the Tristan islanders. Further, we have been instructed not to express any opinion on this problem until the data we have collected have been examined by the relevant authorities.

The next four hectic days in Cape Town culminated in a grand cocktail party to acknowledge the efforts of everyone who had contributed to the success of the venture. Whilst five of the team were returning home on the liner *Windsor Castle*, an advance group consisting of the scientists Ian Gass, Peter Harris, Peter Baker and Donald Baird arrived back in London by air on 4th April. At a press conference Dr Gass highlighted the expedition's key finding that Tristan da Cunha was much more volcanically active than had been previously thought, because the previous eruption (at Stonyhill) had occurred as recently as 200-300 years ago and there was no evidence to suggest with any certainly the time or place of the next volcanic eruption on the island. Asked whether it would be safe for people to return to the island, Dr Gass said, 'It would be a risk, but just how much of a risk we do not know.'

A few days prior to this Allan Crawford had written to the Royal Society requesting that his films be sent back so that he could prepare to give presentations to several organisations such as the South African Navy and Weather Bureau. Responding on behalf of the society, its expeditions officer George Hemmen confirmed that the films would indeed be returned, but he went on to express concern that Crawford had been giving advice and putting forward proposals to various authorities in South Africa about a future resettlement of Tristan da Cunha, and that these had been reported by the British press. The chastening tone continued as Hemmen briskly drew Crawford's attention to the fact that he had agreed in writing to accept the regulations concerning the press and any associated publicity. The rebuke ended with the insistence that any statements regarding the future of the island must originate from the Royal Society, guided by Dr Gass.

Could resettlement ever be contemplated?

It was on Friday 13th April that the *Windsor Castle* docked at Southampton with Adam Swain and Joe Glass on board and soon after they had joined their families in Tristan Close at Calshot. As might well be expected, both men were the centre of attention as the islanders heard directly from them of their experience in the volcano-ravaged village. Their message was clear and positive. The houses had been ransacked but, apart from Dennis and Ada Green's property, all of them were still intact, since they had been spared the lava flows that had engulfed the factory and landing beaches. Adam and Joseph told of the new Garden Gate Beach which offered a landing place partly sheltered by the newly formed lava cliffs, and of the pastures which were in good condition. Therefore, for the first time since that dreadful day when they had evacuated their homes, the islanders realised that there was a chance that they could return. In summary the two men reported, 'There is nothing to stop us going back. Everything is all right,' and that was the message which appeared next day in the press.

Willie Repetto was one of those gathered in Norman Swain's house to welcome Adam home. Eager to voice his feelings to journalists he said, 'I speak for all my people when I say we don't like it here, and we want to go home. We are ready to go back any day and it is only up to the Colonial Office.' Norman Swain added that no one had purchased a television set as they knew they would be going back. Keen fisherman Benny Green, recently married to Sylvia, added, 'Everyone has tried to make us happy, but we do not like it here; it is far too cold. Nothing makes up for our sunshine, and we all want to get back. There are five longboats still on the island, and we have one with us. We could start fishing as soon as we arrived back.'

That same evening Peter Wheeler organised a meeting with the islanders in the chapel at Calshot, so that everyone could hear a report of the Royal Society Expedition from its leader Dr Ian Gass. If the presentation had taken place the previous day the outcome would have been quite different, but the Tristans now approached the gathering with a resolute determination to start planning their return home. Gass spoke of the Tristan volcano and its unpredictability, using a form of words which are totally unacceptable today, when he blithely remarked, 'It's like a woman; you never know what she will do next.' Crucially, he emphasised that it was impossible to say whether the volcano would erupt again and told his audience that the volcano had also erupted at Stonyhill 200 years before. Nevertheless, he expressed the view that the expedition thought the present eruption was dying and 'almost finished', adding that 'I am hoping that someone, possibly myself, will be going back in three to four months. Not until then can we be satisfied that it is finished.' Questions were put to Adam Swain and Joe Glass who continued to be positive about an early return, and almost immediately the mood of the meeting grew to one of 'We want to go back to our island'.

Peter Wheeler then had to confirm the official position that the dilemma surrounding whether the islanders would ever return was 'an open question'. He stated that Gass was writing a report that would be presented to the Colonial Office. It would then be for the Secretary of State for the Colonies to make up his mind, adding, 'Please believe me that neither Dr Gass, Father Jewell nor myself are against your going back.' When Wheeler went on to say that they would have to consider whether it was fair to take small children back to the island, there was an immediate cry from the women, 'When we go back our children go with us.' At this point the meeting got out of control, and Wheeler and Gass left. Father Jewell attempted to calm the situation, but the islanders felt let down and remained angry. Jewell was criticised for failing to pray for the safety of the island and an uncomfortable atmosphere prevailed. Fundamental was the feeling that they were abandoned, and lacked any apparent voice or influence, since they had been told that any decision would be taken by Government officials. The islanders were fully aware that both Wheeler and Jewell were being moved on, and a realisation began to dawn that night that they themselves would need to develop an unaccustomed assertiveness to achieve their wish to return to the island.

In a letter to the Colonial Office on 16th April, Peter Wheeler noted that the material side of the islander's resettlement at Calshot was now almost complete but before he and Rev. Jewell left, all interested parties should be brought together so that the burning question of a possible return could be fully aired and discussed. He proposed that SAIDC, the SPG, Sir Irving Gane and representatives from the South African Government should be invited to join Colonial Office officials at such a meeting. He also recommended that Allan Crawford should be offered the air fare to fly from Cape Town now that he appeared to be heading the campaign for the return of the islanders. If Crawford were brought into the discussions, he would be better able to appreciate all the points both for and against any return to Tristan.

On 18th April whilst awaiting the Royal Society Report, Garth Pettitt and John Kisch from the Colonial Office met to consider Peter Wheeler's letter. Pettitt noted that the islanders had shown their desire to go home but he now favoured a delay in planning, partly as young islanders would become so used to England that they would wish to stay and would consequently persuade others to do the same. His main concern was the islanders' safety. If the report found that there was a risk of further volcanic activity on the island, the Colonial Office could not return the islanders, nor allow them to go back without its authority. Pettitt also thought that SAIDC, headed by Hubert Gaggins, would be eager to have at least some of the islanders back to re-establish a fish processing factory, and would therefore attempt to help the islanders get back, independent of British Government approval.

In the House of Commons, the following day, the MP George Thomson asked the Secretary of State for the Colonies whether he was now able to report on the adequacy of the resettlement arrangements for the Tristan da Cunha Islanders. In reply, Reginald Maudling stated that the islanders were comfortably resettled in homes at Calshot which had been furnished and equipped for them. Over 90 per cent of those seeking employment had been placed in jobs and were receiving the full benefits of local health and social services. The children had been absorbed into local schools within two days of their arrival in Hampshire and had settled down well. Thomson responded by welcoming what had been done to overcome the difficult problems of adaptation here and asking whether the minister had noticed reports that some of the islanders appeared to want to return to Tristan if that were at all possible. Could he give any information on this and would he ensure that everything was done to meet their own desires in the matter? Reginald Maudling replied that, 'I think that it is true that a number of them feel homesick and want to go back. I believe that running into one of the worst winters for a long time has not helped. I am keeping a close eye on the situation.'

By 19th April, the irrepressible Allan Crawford continued feeding his views to the South African press, telling the *Cape Times* that he planned to inform the Colonial Office that it was feasible, and apparently safe, for the islanders to return to their homeland. The *Cape Argus* contacted Dr Gass who responded by saying that 'I am surprised to hear this recommendation from Mr Crawford. I stick to what I have said – to send these people back home is not a risk I would be prepared to take at this stage. There ought to be a re-examination of the situation in four months' time.' Crawford set out his robust views which were published in the *Cape Argus* of 21st April:

> I think that it is possible in time many of the islanders will be able to return to Tristan. I have never suggested that they should return immediately. It could not be done [now] as it is the wrong time of year. What I envisage is that they should return in dribs and drabs over the next few years. By the time they are able to return we ought to have a better idea of whether the volcano is safe.
>
> I think it is desirable for them to return because they are so keen and are quite prepared to take any risks involved. Secondly, their labour is invaluable on the island

both to the important weather station and to the rock lobster fishing industry. I think the volcano was a good thing because there was a population problem on the island and now all the islanders have had the opportunity of seeing the outside world.

In public, at least, matters rested until 14th May when Chief Willie Repetto, with the assistance of his niece, and almost certainly spurred on by Allan Crawford, wrote a powerful letter to Gordon Whitefield in which he outlined his pragmatic view of the urgent and essential work that must be done if the islanders' way of life was to be resumed. This is what he said:

I am writing on behalf of the Tristan people. All of us are most grateful for everything the Colonial Office has done for us. Now we are anxious to know if a date could be fixed for us to return to our homes on the island. If this was done the people would be more settled in their minds. The people are worried that if they stay here in England too long and don't get back to their homes soon the houses will be gone to ruin both inside and out.

They also feel that the five newly built longboats that remain on the island will soon be rotten with no one to paint them or look after them. Another thing is that if they don't get back soon all the trained oxen, which are so important for heavy transport on Tristan, will be going wild; it is impossible to train fully grown oxen and it may be a long time before young ones are trained again.

All the men feel strongly that they must return home in June or July this year. If they don't get back then it will be too late in the season to prepare the ground and plant potatoes, so that there will be no crops for this year, and they will have to wait another year. If there is anything in this letter with which you cannot agree or would like to discuss we should be most grateful if you could come down to Calshot and see the men and talk it over; then we can decide what is best to be done.

In a memo two days later Christopher Eastwood, a civil servant in the Colonial Office, considered the issue of the islanders returning home:

> It seems to me … that we ought to proceed on a basis of purely common sense. The island is clearly volcanic; a serious eruption occurred a few months ago and another one might occur at any time, possibly not for 200 years but also possibly next year. It seems to me obvious that we should be wrong to let the islanders go back. Mr Wheeler and Mr Jewell, the Chaplain, agree with this. I recommend that a firm decision should be made to this effect.

However, the psychiatrist Dr J B Loudon, who had been observing the islanders in England on behalf of the Medical Research Council, advised that it would be dangerous at this stage to say definitely that the islanders should 'never' go back to the island since he anticipated that it would all too likely produce an outbreak of hysteria. In discussion with Dr Loudon and Peter Wheeler it was decided to say quite firmly that there could be no question of going back this season. That approach would gain time and possibly the islanders would change their minds, since a number were settling down well, to the extent of making major purchases such as motor bikes. Several obstacles to a return were noted, including the difficulty of landing and the absence of any company plans for shore-based fishing. It was agreed that Gordon Whitefield would reply to the letter from Chief Willie Repetto inviting four or five islanders to London to follow up the Royal Society report when it was published. If that strategy was followed, the news could be broken to them in London rather than to the whole community in Calshot where emotions were running high and it might provoke widespread distress.

On 21st May Allan Crawford wrote a long letter to Reginald Maudling after receiving requests from various islanders for his help in hastening their return to their island homes. He said that a number were desperately unhappy in their strange surroundings and that 'as a firm act of faith, they are placing their future in the hands of God, convinced that He will lead them back to their own homes'. Crawford stated that the volcano was no longer dangerous and, provided the British Government did not wish to withhold from them their freedom of choice, it may be possible to arrange for the return of some islanders to the best advantage of all concerned.

He went on to build a rationale for returning, these being among some of his more telling points:

The occupation of the island in the name of the Crown was of considerable importance not least because a Russian whaling fleet had recently visited the island, where, as we now know, it had slaughtered many of the breeding population of southern right whales.

The islanders' unique knowledge of the island made them invaluable to employers, currently in the radio and meteorological stations and particularly those involved in fishing. In recent discussions with the company directors, he had been told that they were keen to employ island men aboard their fishing boats since they were superior to any labour from the Cape. As the new fishing season was due to start in August and provided the British Government did not start to restrain the islanders from leading the life of their choice, a selected number should be permitted to take part.

He proposed that twelve able-bodied and industrious Tristan men be selected forthwith by the community and that they should proceed to Cape Town in August. The group should be led by someone like Thomas Glass, accompanied by another who had a good knowledge of the damaged water supply system. Six should be employed by the fishing company to work aboard the fishing vessel *Tristania*, coming ashore when the ship departed for Cape Town, while six should be landed and prepare the island for the later return of the community. These last would be involved in such tasks as repairing houses, planting potatoes, restoring the water supply, and repairing boats.

If by Christmas they were satisfied that the island was fit for the return of their families, the relatives of that advance party could be sent out from England to South Africa awaiting onward travel to Tristan. This leg of their journey could possibly be made aboard the *Tristania*, the South African Department of Transport's new research vessel, the *RSA*, or HMS *Protector*. After the re-establishment of a small community, it would be possible to consider re-opening the radio and weather stations and rebuilding the factory.

Crawford envisaged a planned gradual repatriation to enable the return of those who desperately want to do so during the summer of 1963/64. As others became more settled in England, they would remain, and so Tristan's problem of the last twenty years – that of over-population – would have been solved for a considerable time to come. He suggested that in due course the British Government would decide what amenities it would be prepared to provide (if any), such as an Administrator, a doctor, a schoolteacher, or an agriculturalist. If this expenditure should be unwarranted, the SPG might be able to select a clergyman and his wife who could run the school. Here then was a report of considerable weight, and in order to compile it Crawford had turned in discussion to such prominent figures as the British Ambassador to South Africa Sir John Maud, the Archdeacon of Cape Town Roy Cowdry, and Sir Nicholas Copeman who was the Commander-in-Chief of the Royal Navy's South Atlantic and South America Station.

Crawford adopted different positions before and after this letter of 21st May 1962, but it was his pragmatic idea of a team of twelve islanders to return later that year, six to prepare the village, and six to fish, which proved to be the key to unlocking a successful return for the refugees and its significance was fundamentally important as it proved a blueprint that the British Government could not ignore. Nevertheless, there would be further twists to the story before Crawford's plan was put into operation.

On 31st May a deputation of seven Tristan islanders visited the Colonial Office in London to meet Christopher Eastwood. His subsequent internal memo summarised the discussion in this way:

I told them that the Secretary of State had given much thought to the question of whether or not it would be possible for them to go back to Tristan. Two of the factors were the risk of further eruptions and the question of the beaches. I understood that that they were prepared to take the risk of another eruption, but the Secretary of State was very doubtful whether, since the eruption, the beaches would be good enough for it to be safe for anyone to be put on the island. The Secretary of State had therefore decided that he could not undertake the responsibility of letting anyone go back until after the Tristan winter. He hoped to be able to arrange another survey of the island, by HMS *Protector* or another naval vessel.

Eastwood was aware that the news would be a disappointment to the islanders since they were anxious about their houses, the planting of potatoes, and the fact that their oxen were going wild. The islanders took the news fairly well, but some were disappointed at the delay in making a further inspection and advocated that it should take place in July or August, which might possibly enable them to return in October or November. The matter of what supplies, including food and building materials, they would need was also discussed. Eastwood floated the idea of an advance party of a few able-bodied men (as suggested by Allan Crawford), but this was rejected by the islanders since they all wanted to stick together. Eastwood gained the impression that they were all perfectly sincere and convinced in their desire to return to their island, but Peter Wheeler, by now a holder of the OBE for his work as Administrator of Tristan da Cunha, assured him that 'fully half of the community do not in their heart of hearts really want to go back'. He advised strongly that, rather than have an elaborate Government-backed repatriation plan, it was best to let them return independently (subject to assistance in finding a boat to transport them) when they had saved up enough money to do so. He thought that if they all went back, many would miss the higher wages they had been earning in the UK and within a few months they would want to return to England. In a cautious mood Eastwood suggested the thing to do was to play for time, with the next move being to try to arrange a second inspection of the volcano later on in the year.

In response to that meeting with the Tristan deputation, the Colonial Under-Secretary Hugh Fraser felt a concern that the islanders had been left with the impression that there was a strong possibility of their returning to their island. Therefore, in the following few weeks, he urged doing everything to remove these false hopes. They would only make it even more difficult, he argued, for the islanders to take the likely refusal to let them go home after the planned inspection. So, there were many conflicting currents: the letters from Allan Crawford; the line agreed to be taken with islanders; and Hugh Fraser's guarded concerns about whether a second survey should proceed later in the year. Christopher Eastwood reflected with, we must suppose, a heavy heart, 'though it is obviously inexpedient to say this to them at the moment, I remain of the opinion that it would be better that they should never go back.'

In what may have been his final correspondence on Tristan matters before he was moved back to Kenya by the Colonial Office, Peter Wheeler wrote a thoughtful letter to Allan Crawford on 11th June. The content is significant because of the way Wheeler expressed his own private view on the situation. In it he said that he was not keen, despite the arguments that their houses and boats were deteriorating, that any firm decisions should be made to send the islanders back 'as this will undoubtedly jeopardise the chances of them settling down happily in this country'. He thought that while the islanders said publicly that they were anxious to return, in private they indicated (rather than stated) that they would prefer to remain in England, adding that 'the glue that binds the community together is still so strong that none of them is prepared to stand out and say he is not anxious to return, but I feel that given the chance of settling down in this country, perhaps 50 per cent would come out into the open and make it known that they would definitely prefer to remain'.

Wheeler confirmed that he, as well as Crawford, thought the island was over-populated before the evacuation, and this was another reason why 'every opportunity should be given to those who wish to remain to settle down and get both feet on the ground'. Reaffirming his view that no firm decision on return or non-return be made at this moment, Wheeler was aware that many would never be able to forget their island home and would find it very difficult to settle down in this country. 'You will know who they are, and I hope – although this is completely unofficial, and I am talking from a personal point of view – that they get the opportunity of returning in due course.' Reflecting on the question of whether it would be good for their sea passages to be paid by the Government, he pointed out that the islanders were now earning decent wages, and therefore quite capable of saving enough money to pay for their own tickets. Referring to the help that the islanders had received in Britain, Wheeler mentioned the numerous hand-outs of clothing, equipment and even money. He had 'noticed a marked deterioration in attitude towards this over the months', and concluded with these reflections:

> There is a definite feeling amongst some that it is the duty of the rest of the world to look after them and provide for them in every respect. This, of course, is bad for everyone, and I think if only to gain their self-respect, they must begin to stand on their own feet once again. I want you to understand that I am not against return for anyone who so wishes, but I am very much against spoon-feeding, which is inclined to sap the morale of any of us in time.

Others involved in these continuing debates naturally had opinions too. One of those present at a meeting at the Colonial Office at the end of May was the SAIDC representative Maurice Willis. (himself a former Colonial Office official). Within a fortnight he wrote to Allan Crawford, describing how the older islanders were very despondent and homesick, but adding that it was difficult to be sure how the younger generation felt. They had, he said, reached the stage of undertaking to pay for their own passages back to Cape Town. Commenting on Crawford's long letter of mid-May he thought the future of the islanders 'will be built on sand as the people are too unstable'. He pictured a scenario in which the view of the older members of the community would prevail, thereby influencing the younger men to return, but the latter, having had experience of a more exciting life, might well drift away from the island, leaving the elders to fend for themselves. He thought the company could not be expected to invest in building a factory without a reasonable chance of co-operation from a settled population, and that must include the young fishermen.

On 20th June Willis wrote to the Colonial Office clarifying his company's position. He was disturbed by the rising expectations amongst the islanders that his employers would provide significant assistance in the resettlement of Tristan da Cunha. The company would be willing to employ islanders as fishermen or deck hands aboard its fishing vessels, and these youngsters could be the nucleus for resettling the island. Nevertheless, it would be out of the question, after the heavy financial loss suffered as a result of the volcanic eruption and the resulting failure of the old company, for the new enterprise to invest fresh capital on the island or to accept any commitment to the islanders involving expenditure. In passing, it is telling to reflect on the sensitivity of these discussions. As an example, when Rev. Jack Jewell was forming plans to write a book about Tristan and approached the Colonial Office with a request for copies of official telegram messages between that department and Peter Wheeler, his application was refused, quoting in justification the 50-year embargo on access to official documents. And only a few days earlier, in a parting shot, Wheeler hazarded the thought that it would be a most difficult job for any manager to handle men who had developed dangerous ideas from their sojourn in trades-union conscious England. Furthermore, the company could not be left with the responsibility for transporting islanders back to Cape Town if they became dissatisfied with life on Tristan or if conditions there had deteriorated. And tellingly, there had been no official indication from the South African Government that it had any wish to re-establish the meteorological station on Tristan.

Islanders exercise their right to choose

The islanders were developing a new life for themselves. Several young Tristan women met future husbands on nights out in the local area and settled in Hampshire. And yet blending-in continued to prove a challenge with medical and social scientists and the media still eager to follow their every activity. There were frustrations too. Exasperated that he had received no communication after their meeting of 31st May 1962 Chief Willie Repetto wrote to the Colonial Office, asking for twelve men to be given the chance to travel to Tristan in August 1962. The letter was an ultimatum of intent signed by 67 of the island men, and it read:

If the Colonial Office isn't going to make any arrangement by 1st August about sending the men out to Tristan, the men is going to pay six men passage themselves over to Cape Town. What is the use of sending people out in October when the winter will be over, and the houses will be gone to ruin, and our oxen will all be all gone astray? It is no good sending the expedition out to look at the island without islanders go as well because the islanders know more about the landing beaches. I would like the answer back as quick as possible. The names on these pages are the men which are going to pay for the six men passage to Cape Town.

It undeniably had some impact. In South Africa the following week the *Cape Times* reported Willie Repetto's ultimatum. It is thought the letter was probably drafted by Allan Crawford with a view to putting pressure on the Colonial Office for his proposed scheme. The next day the *Cape Argus* stated that the islanders' hope of travelling home with official help had faded away when they had been advised that the UK Government had no plans for them to return that year. On 17th July Rev. Jewell, described as the 'former chaplain of Tristan' working elsewhere, visited the Colonial Office to enquire about the basis of various reports in the press concerning a return by islanders that autumn 'off their own bat'. He was told that the campaign had been master-minded by Allan Crawford and that Colonial Office attempts to stop Crawford 'meddling' had been quite unsuccessful. The officials suggested that Sir John Maud, as the country's ambassador in South Africa, might be able to persuade him otherwise. Jewell thought that, if the Government did not facilitate it, the islanders would not have the resources to go back under their own steam. He had always been opposed to the idea of any return to Tristan because, in his view, those going back would never settle and, in any case, they did not know what they were going back to.

That same afternoon Father Redmayne of the SPG visited the Colonial Office and was briefed on its position. He was strongly opposed to the islanders returning, but if the Government did decide to proceed with the plan the SPG would continue their interest as before. He informed them that the sizeable sum of £17,000 had been collected to provide a new roof for St Mary's Church on Tristan but, if the islanders did not return, it was agreed that this fund would be used to purchase and renovate the chapel they had been using at Calshot, which would then be placed under the Diocese of Winchester.

Christopher Eastwood took time out to mull over the implications of Willie Repetto's demand. On 19th July, in answer to his own question 'What attitude do we take to this request?', he made the following observations:

Firstly, the Colonial Office had received one private letter from an islander disagreeing with the proposal, the first sign of opposition to the project amongst the community.

Secondly, he could not see, if the islanders were able to pay their own passage to Cape Town, how the Government could stop them returning. Apparently, most of the community would subscribe to the cost of these journeys. He also flagged up the possibility of finding money from the existing welfare fund.

Thirdly, the strong advice received from those who knew the islanders best (such as Peter Wheeler, Jack Jewell) was that the islanders should not be encouraged to return to the island.

Fourthly, he recommended that the official policy should be to continue to be as discouraging as possible. But if the islanders insisted on going ahead with the plan for a dozen of their men to travel out the following month, his department should not attempt to prevent it, and indeed should try to find a little money to help with their passages, but it should be made clear that this carried no implication at all that plans would be made for any other members of the community to return home.

Finally, reassurances should be given that the Tristan community would be able to live indefinitely in the houses at Calshot as long as they continued to pay rent.

On 21st July Chief Willie Repetto arranged a meeting of the island men on the open square at Tristan Close. This gathering, called by the islanders themselves and without any Government official present, was the first such assembly since the times back on Tristan when there were no priests, naval commanders, or UK-appointed Administrators on the island. Highly significant was the fact that the islanders were now fully resident within their local area and had no outsider to lead them. In marked contrast, it was left to their own initiative to attempt to liaise directly with officials from the Colonial Office. This was the day that Willie Repetto put the idea of the Resettlement Survey Party into practice. He spoke in a booming voice, which could be heard by all present, underlining the seriousness of their situation and the lack of support from officialdom. With great drama he declared, 'This is our last chance to go back to our homes.' He wanted to know if anyone wished to remain in England but added, not without a hint of manipulation, that he would keep to himself what he thought of them. He then asked for twelve volunteers to make the journey back, six to join the fishing ships, and six to work on the island, whereupon he actually named them one by one, and as each man was selected, he stepped forward and stood before the crowd.

His nominees were the fishermen Albert Glass, Martin Repetto, Stephen Glass, Benny Green, Anthony Rogers and Leonard Glass; and the shore-based party Thomas Glass, Johnnie Repetto, Lars Repetto, Walter Swain, Neville Glass and Philip Green. Willie Repetto realised that every one of the menfolk wanted to be among the first back, but he was careful to choose men with no dependent children, and to make sure the party was not dominated by his own close relatives. Of the twelve, five were married but had no children, and the others were single. The six who would work on shore would be led by Willie's brother, Johnnie Repetto, and Thomas Glass, both aged 50. These two would assess the state of the recently abandoned island and answer the formidable question: was Tristan still unsafe, in which case everyone would have to come back to Britain; or was it habitable once again, in which case they would recommend a full resettlement.

Discussion followed and it was agreed that the families of the party members would be supported by contributions from the other islanders who were working locally, as they would lose the income that the absent men would otherwise bring in. This was accepted on the assumption that the twelve men would repair and look after all the houses on Tristan when they arrived. There were believed to be a couple of island men who were not so enthusiastic about returning.

The psychiatrist Dr Loudon thought overall that morale was remarkably high considering the traumatic turmoil that the evacuees had experienced. A few families were secretly hoping they would not have to return to Tristan, but they were in a minority, and kept such thoughts to themselves. The refugees' health was good and had appreciably improved since they had arrived in Britain. Even though cash had been saved to purchase tickets, it was hoped that SAIDC would reimburse the fares of the six fishermen who were destined to work aboard MV *Tristania*.

On 23rd July, a deputation of islanders travelled to Southampton with about £500 saved by the community to book tickets for sailing of the *Stirling Castle* to South Africa departing on 9th August. On their arrival at the Union Castle Line offices, they were told that the ship was fully booked. This was the last departure that would enable the party to catch the outward passage of the *Tristania* to Tristan at the end of August. This was a severe blow to the men, but fortunately a change of policy was beginning to emerge in Whitehall.

On 26th July John Kisch and Garth Pettitt from the Colonial Office and Maurice Willis from SAIDC travelled to Calshot to meet Chief Willie and ten other men. Willis told the islanders that they could not count on a new shore-based factory being built, so the employment for men and women that they had become used to would not be available on their return. For their part, the islanders restated their position that nothing would shift them from sending a party of twelve the following month, as this was their last chance, and they were quite unwilling to wait for the results of any proposed survey in October. However, they admitted that if the beaches were unusable, resettlement would be impossible. They insisted that they were the only people who could make this decision after inspecting the shores of the island themselves.

In a subsequent memo summarising the meeting Kisch, Pettitt and Willis stated, 'There is little doubt about this as the return has become an obsession even with the small children.' They were quite clear that there was no way of preventing the pioneer party from going and they therefore recommended that the Colonial Office cooperated to ensure that the best possible arrangements were made. Whether this would lead to a large-scale resettlement depended largely on the state of the beaches, 'but it might be [a possibility], more especially since it appears likely that in a year or two the company will wish to establish the fishing factory'. Willis had little doubt that the lobster industry on Tristan would be profitable.

At last, there was a shift of mood. Before the end of the month the Colonial Office had contacted the Union Castle Line to enquire whether twelve tourist berths could after all be found on the *Stirling Castle*. In reply the company confirmed that twelve spaces had been reserved at tourist-class prices, by switching some existing bookings to first-class accommodation since all the remaining tourist berths had been taken. At the same time the Colonial Office approached Sir Irving Gane, explaining the islanders' plans and the need to find £1000 for the party's outward passage. A contingency of £1000 was also sought to pay for their return at Christmas if conditions on Tristan prevented the resettlement of the rest of the community. Gane questioned why the Tristan da Cunha Fund should be used to fund the six members of the fishing party, since the fishing company should pay their fares if they were wanting them to work on their ship. Nevertheless, he agreed that the fund would pick up the £1203 bill for the entire group, notwithstanding that the company should be asked to reimburse the fares of the six fishermen or at very least make a substantial contribution towards them.

On 28th July Allan Crawford wrote to Willie Repetto with the news that SAIDC had agreed to advance £50 on the six fishermen's wages to pay their fares on condition that they sign on for the season. The sum would then be deducted from their earnings at the end of the period. He also warned them that he was quite certain that the Government was not going to arrange for the full resettlement of the island as easily as they thought. A return for the remainder of the community would undoubtedly be very expensive, since an unusually configured ship would have to be exclusively chartered to carry not only its many passengers but also the mass of supplies, including sufficient food to sustain them for at least six months.

On 1st August four islanders, Willie Repetto, Johnnie Repetto, Thomas Glass and Lars Repetto, attended a meeting in London with Nigel Fisher, Under-Secretary at the Colonial Office. During the discussion Fisher said that the proposed return of the men must be regarded as a trial trip for the purpose of ascertaining whether the beaches were useable, and that the question of resettlement was a quite separate topic for consideration. Whilst offering to facilitate the islanders' undertaking, he imposed a further condition: that a representative of the Colonial Office, probably the former Agriculture Officer and member of the Royal Society Expedition, Dennis Simpson, should support the expedition in order to prepare a report to the Colonial Office on the feasibility of resettlement. (In the event, Simpson was unable to join the party, and so the Colonial Office engaged a former Agriculture Officer, Gerry Stableford, to take his place.) After the meeting, the islanders were interviewed by a number of journalists. In answer to one of the questions Lars Repetto said in a slightly gung-ho fashion, 'We are longing to get back.... We do not like television at all. We haven't found any agricultural ideas that would suit us to take back; in fact, we had plenty of food on the island.'

Among the series of journalists who visited the islanders in Calshot at around this time was the Danish travel writer and photographer Arne Falk-Rønne. As was the custom, it was 60-year-old Willie Repetto who received him with his usual courtesy. Falk-Rønne later recalled the conversation with Willie who told him that the community was grateful for the help they had received in Britain, but they weren't happy in a high-paced covetous society more intent on gaining wealth and were missing peace and quiet and the wonders of life – the sea, the mountains, the fresh air.

'We've had a meeting about it, Willie continued, 'and almost all of us are determined to go back. For about 150 years we and our forefathers have inhabited Tristan da Cunha, and in all those years we have only been seven families, the same seven families as were evacuated when the volcano erupted. Until now, some have felt that it was better – and safer – to remain in England, but since Mr Gass has returned, we are sure we shall go back.'

When Falk-Rønne pointed out that Dr Gass's report was not particularly optimistic, Willie replied that it was optimistic on one very essential point. He said the lava had avoided destroying any houses in the settlement, and that cannot be other than God's work. 'By this He has shown us that we can and shall take up our existence again on our forefathers' island. From now on, *nothing can stop us.*'

Arne Falk-Rønne planned to travel to Tristan with his compatriot Peter Anders Juhl, and so he spoke to the Resettlement Party's leader, Johnnie Repetto and asked him to agree that he would come out to any ship that was to anchor off Tristan and take him and Juhl ashore, as he hoped they might be the first true 'visitors' to the island since the eruption.

The party had a setback only two days before they were due to sail when they heard that the fishing company stipulated for the first time that the six fishermen had to give a firm an undertaking to work on the ships for the whole season lasting a full six months. The men were taken aback by this request and agreed that no one could stand six months' fishing, yet they had no option but to make such a commitment since all the passages were already booked. The following day, the company's UK representative Maurice Willis met the island's previous Administrator Peter Day. They both shared concern about the long fishing commitment and doubted whether the other six men could do all that was necessary for resettlement. They also expressed doubts about the sheer practicability of uplifting the whole community with their belongings from Calshot, and conveying all its members back to Tristan, even if a report for return was favourable. Hardly an optimistic prognosis on the eve of the expedition.

The survey party sets off

At 4pm on Thursday 9th August 1962, the twelve men of the Resettlement Survey Party left Southampton Docks aboard the *Stirling Castle* bound for Cape Town. The following day Maurice Willis confirmed to Allan Crawford the departure of the group and conveyed his anxieties in the following terms:

> We all have the real welfare of the people at heart and we are desperately afraid of them getting into a mess. You have let yourself in for a great responsibility as the people look to you and there is really no one else in South Africa sufficiently interested to deal with any eventualities that may arise. I wish you luck. We have all urged the more responsible men amongst the advance landing party not to give a favourable report unless they feel certain that everything is all right for resettlement.
>
> [The Agriculture Officer] Stableford will be some check on that, but he will be coming off the island again in October and won't have a lot of time to do a proper survey. With your intimate knowledge of the island, I hope that you will be able to ensure that the men are giving us the right picture. It isn't enough to say that because a small party of able-bodied men can scramble ashore, the old and women and children with all their worldly goods can do likewise with safety.
>
> I beg you not to let your desire to see the people happy sway you from reality if you feel that, notwithstanding what the advance party may say, there is danger of a disaster. I am sorry to write like this, but we do feel worried.

The *Stirling Castle* arrived in Cape Town at 10.30am on 23rd August. Interviewed by a *Cape Argus* journalist, Johnnie Repetto said,

> We don't know how long we will take to decide whether our island is safe for everybody to come back to. The biggest thing to decide is whether there are landing beaches for our boats. If there are, then I think we will be able to stay on Tristan. The volcano is dead. Even if it is a little lively, we aren't afraid of it. When the time comes for the young people to decide if they are to stay in Britain or come back, I think they will all come. We have no grumbles about Britain. Everybody has been very kind to us. We are like the Israelites. We are chased out of our land and like them, we sat down and prayed to be allowed to return.

Albert Glass added, 'All the young people will come back. There is more to get out of life on the island. We are glad to be here – but we'll be happier when we reach Tristan.' As the *Stirling Castle* docked the twelve islanders met Gerry Stableford, who had flown on ahead of them, and they were joined by Allan Crawford and Bishop Cowdry.

Crawford received a pithy briefing from John Kisch in which he confirmed that the British Government had resolved to assist the exploratory visit of the islanders, adding that it had been brought home extremely clearly that the Colonial Office needed to

Members of the Resettlement Survey Party board MV *Stirling Castle* at Southampton Docks on 9th August 1962 waved off by their families. (Tristan Photo Portfolio)

make all arrangements, in agreement with the islanders, since they were quite unable to make even simple administrative arrangements themselves. It therefore followed that the British Government would need to take full responsibility for anything that was later required, such as the employment of teachers and doctors. He added that, whilst supporting the present expedition, the Government was not committing itself to a decision on the resettlement of families, and the islanders had agreed on this point. Crucial would be the condition of the beaches and the intentions of the company, who had seemed inclined in principle to re-establishing the lobster industry on pre-volcano lines but had not yet made a commitment about building a new factory. If there was no fishing factory or shore-based fishing, the return to a subsistence economy which relied on farming and fishing for home consumption would not be such an attractive one for a people who were used to higher wages and the other advantages of living in England.

Because Gerry Stableford arrived in South Africa in advance of the others, he had been able to co-ordinate extensive preparations in time for the departure of the Resettlement Survey Party. SAIDC had now agreed to employ the six fishermen for the first half of the season but would bring them back direct to Cape Town from Gough rather than dropping them off at Tristan. (In the event the *Tristania* did indeed take the men back to Tristan before returning to Cape Town for Christmas.) Stableford organised the purchase of enough stores including food to last the seven-man land-based team five weeks until the visit of HMS *Puma* in October. He met Vice-Admiral Sir Nicholas Copeland, who provided 'walkie-talkie' radios to communicate with the *Tristania* and paid a visit to the Mayor of Cape Town who helped with a stock of medical supplies. However, despite great interest in the expedition, Stableford did not want to accept more charity from the generous people of Cape Town. The South African Government was preparing the new Antarctic supply ship *RSA* to take provisions and re-fit the Gough Island Weather Station and in principle the South African Weather Bureau was interested in re-establishing a meteorological station on Tristan.

Perhaps Stableford's most vital task was to retrieve the island's firearms. Those secured by Peter Wheeler in October 1961 had been handed to Scotty. Undeclared in Cape Town, they were seized by South African customs officials when they searched *Tristania.* Stableford was able to persuade customs officers in Cape Town to release

Members of the Resettlement Survey Party gathered aboard MV *Tristania* in Cape Town before their departure for Tristan on 30th August 1962. Left to right: Benny Green, Johnny Repetto, Martin Repetto, Thomas Glass (squatting in front), Philip Green, Leonard Glass (top of head behind) Anthony Rogers, Lars Repetto, Walter Swain and Albert Glass. Neither Stephen Glass or Neville Glass are visible. (Allan Crawford)

the twelve rifles and ammunition from the Customs Bond Warehouse to take back to Tristan. This diplomatic coup demonstrated the co-operation and goodwill of the South African Government authorities towards Tristan da Cunha even after the country had withdrawn from the British Commonwealth. Rifles were used on Tristan to shoot the feral dogs, wild cattle at Stony Beach, and to carry as a defence against shark attacks on longboat trips to Nightingale or Inaccessible. Now all was as ready as it could be and on Thursday 30th August the *Tristania* set forth to carry the Resettlement Survey Party out to the archipelago.

The Survey Party get to work

Tristan was sighted at 8am on 8th September 1962 and the *Tristania* finally arrived close to the island at 12.30pm. The day was mercifully sunny and clear with a light WSW wind and sea flat calm at Garden Gate. This meant that the party of islanders and Gerry Stableford, together with their 5 tonnes of gear, were safely and speedily unloaded by 2pm using three of the vessel's fishing dinghies. The new Garden Gate Beach had formed over the original Little Beach rock pools and consisted of small- and medium-sized stones. It was workable with small craft, but the party realised that hauling up loaded longboats would not be feasible since there were no facilities for

St Mary's Anglican Church in the 1980s. (Richard Grundy)

offloading heavy freight, as there was no sand or a winch, both of which were essential for such an operation. At 4pm the Survey Party arrived at the Residency, still referred to as 'Mr Wheeler's house', to set up their headquarters. For the first ten days all twelve men worked closely together before the six fishermen joined *Tristania*.

As the next day was Sunday, the team began their first full day ashore with a short service in St Mary's Church, led by Lars Repetto, as they would do every week. They then hoisted a Union Jack at the flagpole, caught two donkeys, yoked two pairs of bullocks into carts, and used the pack animals to bring the remainder of their luggage up from the bull pen by the Garden Gate Road. A start was made clearing and tidying the abandoned island homes which were in a state of utter chaos despite the careful work by the crew from HMS *Leopard* and the reassuring reports made during the Royal Society Expedition. Looting was evident, and hundreds of empty liquor bottles were found strewn around the approach to Little Beach. Blame for the mess was put down to visits by crew members from the Russian whaling fleet who may have come ashore, and it was hoped that those of the *Frances Repetto* and *Tristania* were not the ones responsible. The Residency base was found to be in a very untidy state, but the refrigerator was workable, and most items of furniture, including the Island Council tables from the Prince Philip Hall, were intact. Elsewhere the Post Office door had been forced, and postal equipment had been damaged, much of which was now unusable.

Re-instating the water supply was a priority. The water-table level at the catchment spring had dropped, so a further dam was constructed with mud and stones which in the space of just two days raised the water to its usual height and so it flowed freely through the main supply system once again. Some of the water pipes had been fractured, buckled and lifted during the eruption and these would need repairing later. The sewage system seemed to be working but would need attention before any full resettlement. The hospital in the old naval station buildings was cleaned up and set in good order. The men heard dogs barking in the night and, determined as a priority to shoot all of those remaining that they were able to, since the animals continued to destroy the island's livestock.

On 14th September work started on preparing for a crop of potatoes by spading (that is to say, digging the ground and clearing weeds) on three large patches. Three days later in beautiful weather Martin Repetto, Anthony Rogers and Leonard Glass went out in a dinghy and caught 35 bluefish with handlines in just two hours. Many lobster were seen around the rockpools close to the lava field, and five finger fish were easily caught from Garden Gate Beach. This was heartening news: the eruption appeared to have had no obvious effect on local fish stocks. On the following day the Island Store was cleaned out and an inventory was created. Even though its doors and windows had been smashed, tinned food was still usable, although rats had feasted on the sacks of rice, flour and sugar, which meant that these had to be dumped. *Tristania* returned and the six island fishermen, who had provided valuable support in the early stages of the village clear-up, went aboard to start their fishing work aboard as planned.

On 19th September the Dutch cargo boat *Straat Bali* arrived at Tristan, nineteen days after leaving the Indian Ocean island of Mauritius bringing Arne Falk-Rønne and Peter Juhl. The ship's captain refused to launch a boat to take the pair ashore, fearing the high sea and the fact that the usual landing beaches had been destroyed by lava. Falk-Rønne thought the new volcano resembled a 'devil's cauldron, with smoke billowing from several craters … a cheerless lunar landscape'. At first there was no sign the survey team had arrived, but the ship's first mate spotted five men launching a dinghy which was rowed out in the swell and, with the skill for which the islanders have become famous, the tiny vessel soon managed to come alongside. One of them was the party's leader Johnnie Repetto who, recognising Falk-Rønne, welcomed the pair to Tristan. The dinghy needed two trips to bring ashore both of the men, as well as their equipment, luggage, and a supply of provisions.

When Falk-Rønne landed at the Garden Gate boulder beach, Johnnie shook the seawater off himself and said with a smile, 'There's a gentleman who wants to welcome you,' whereupon he led Falk-Rønne along to the rock pools to view a huge bull elephant seal. Tristan humour was alive and well! Soon the second dinghy trip was made with Peter Juhl and as much luggage as could be brought ashore. Therefore, by 12.30pm the onshore party had swelled to nine: Gerry Stableford, referred to as the Administrator, the six islanders and the two Danes, who Lars Repetto called the 'photograph men'. The day showed up a weakness in the size of the shore-based team: six hands was not enough to crew a traditional island longboat with any safety, even though these boats were available and stowed on the plateau above Garden Gate Beach.

The Danes were aware of their isolation and unsure how they would secure a passage home. This was summed up by the islanders saying that 'God decided the day you should land on Tristan, and He alone will choose the day you leave it again'. They observed the lava's dead world of black-burnt stone, and the fact that only one house belonging to Dennis and Ada Green had been damaged, as the flax-thatched roof had been burnt off. On closer inspection they discovered here a cooking pot still standing on the stove, as well as spoons, forks and knives set out on the table for a meal that was never eaten. Young Valerie Green's pram was found twisted and rusty. The door was burnt, the rust-covered key lying between layers of charcoal. Nevertheless, the house walls were intact and had been saved from any damage from the lava which by good fortune had flowed eastwards away from the village. The visitors were told that as their homes and belongings had been spared it was God's will that they should return. They felt because of this, that no harm would come to them when they settled on their island once more.

A few metres away from this house an old-fashioned wind-up gramophone was found, complete with a record on the turntable. Soon it was started up again and the song 'Don't fence me in….' blared out over the deserted cottages. The island men went around all the recently constructed water-closets, each placed to the east of every house. The Danes described them as like a row of sentry-boxes – symbols of modern civilisation – all with open doors, all with white bowls shining in the pale spring sunshine. As the team checked the water supply, they pulled the chains in all the outside toilets, and thus a familiar flushing noise was added to the sound of the surf in the distance.

Falk-Rønne and Juhl entered the church by a door which creaked on its hinges. The wind whistled through the entrance hall and blew the loose pages from a hymn book. An altar cloth had partly dropped to the floor, but there stood a pair of pewter candlesticks with new candles in them, just waiting to be lit for a future service. In one of the houses, they saw driftwood floorboards through which rats had eaten large holes. An old sea-chest with a squeaky lid was opened to reveal linen, knitwear and a Danish ten-krone banknote, one that had been issued in 1945 and was now no longer legal tender. On a small table in the doctor's house there stood a half-empty bottle of port and four glasses. The remains of a cat were also found there, and alongside the Prince Philip Hall a donkey carcass, most probably killed by the dogs.

The Danes settled in to their first evening ashore, observing the island men: Lars cleaning a gun ready to shoot the wild dogs; Neville sorting a fishing line; Walter repairing an old pipe; two others playing Chinese marbles. Under the light of a paraffin lamp an old cowboy song was playing out on another vintage gramophone, grinding away far too slowly (and therefore sounding far too deep). At that point, a strange noise caught the ears of the two men from Denmark, which they thought was a motor horn. Thomas Glass identified it as the call of a southern right whale and went on to tell their visitors about their affection for the huge mammals which used to return every year to their Tristan breeding grounds. They had learned to recognise faithful pairs of breeding whales and had observed three years earlier an unusual calf with two big white spots between its fins. Even though she was only six months old, she was already estimated to be 10 metres long, and they had given her the name of 'Little Spotted Mary'.

The days that followed were packed with activity. The expat houses, though still water-tight, were in considerable disarray and needed to be tidied, their contents sorted and, where necessary, destroyed. The South African Government Weather Station was checked and put into order. Tools had been pilfered, four radio transmitters and other wireless apparatus had been ripped out, though the equipment had generally remained intact. Work continued at the hospital, removing the scattered stores, and turning it from its filthy state into a workable facility once again. Whilst all this was happening, Lars Repetto took the two Danes behind the new volcanic cone so that they could take photographs.

On Sunday 23rd September Thomas Glass, Lars Repetto and Peter Juhl went out to the Molly Gulch to collect Albert Glass's dinghy, which they pulled along Hardie Beach and shoved off at Hardie Cave. Since the boat had a couple of holes in it, they plugged the gaps with handkerchiefs and rowed it back to Garden Gate Beach. The following day Neville Glass and Philip Green collected bullocks to yoke up to carts to transport wood for fencing. When they got to the flax garden, they saw a few of the feral dogs, so they alerted the others and began a chase with rifles which took them down the valley of the Big Watron and along the beach. An ambush was planned but shots missed their target and the dogs escaped over Hillpiece towards the Patches. More guile would be needed to succeed in this essential mission.

On 28th September, the new volcanic cone was quiet, but smoking slightly from the crater and parts of the lava field. Weathering was occurring at the crater where the softer ash was opening out and falling away. Fumaroles, giving off sulphur fumes, were still active in the crater area. Grass was now flourishing right up to the base of the cone and above the crater, where ferns were growing again after being burned off by the eruption. Most of the lava field was cold and the crater was becoming cooler. The lava coastline had eroded landwards during the winter. As well as the boulder beach at Garden Gate, there were new boulder beaches with some sand in the centre of the lava field, and east of the lava, stretching 200 metres to Pigbite Ridge. Both beaches could be used for landing if roadways could be made from the village.

Gerry Stableford served as Agriculture Officer on Tristan between 1953 and 1956, so his assessment of the island's livestock and agriculture was particularly helpful. He was dismayed that the sawmill equipment he had purchased to crop the Sandy Point pine trees was found intact in the island tractor garage as no-one had bothered to use it. No geese or poultry had survived the continued onslaught from the wild dogs. Only fifteen sheep survived dog attacks, but two hundred cattle were found in good condition. There was no adult bull, as the Hereford stud bull had been killed for food by the Royal Society Expedition. Forty donkeys in good condition were found in the village area, and some were caught for use as pack animals. Several cat carcasses were found, and two cats were discovered in a reasonable though 'rather wild' condition. It was thought other cats may have taken to the hills. Whilst rats and mice had ravaged the Island Store and some houses, they were not as numerous as expected nor had they done as much damage as might have been anticipated, possibly owing to attention from feral dogs and cats. Stableford emphasised the importance of destroying all the dogs to ensure the future safety of the livestock. He discovered that the old native potato stocks had been destroyed or died out, which was regrettable since they had built up their own immunity to wind damage and Tristan conditions, qualities that imported varieties would take years to acquire. Therefore, he would recommend importing fresh seed-potatoes from certified, disease-free stock.

Meanwhile in South Africa, HMS *Puma* sailed from Simonstown on the morning of 27th September 1962 under the command of Capt. Mellis for a mission code-named Operation Atlantic Uplift. Radio contact with the Resettlement Survey Party via *Tristania* had been planned, but none had been successfully achieved, and so no orders for stores were received. Nevertheless, the vessel carried a range of articles which were thought might be useful, and these including picks, shovels, cement, paraffin, ladders for cliff scaling, a derrick, a mooring buoy, and demolition materials.

HMS *Puma* arrived at Tristan at 4.30pm on 1st October. The ship anchored close to the *Tristania* to which mail and stores for the fishing vessels were transferred, and in return the naval crew received frozen and live lobster. The ship was available as a contingency passage to Cape Town for the six members of the shore-based Survey Party in the event of their mission proving too dangerous. As it happened, all was well, and so they stayed on Tristan. The six island fishermen aboard the *Tristania* came ashore to pick up their mail and, along with those already on the island, caught up on news from their families. Gerry Stableford went aboard as planned with his luggage and made a preliminary report to Capt. Mellis stating that the islanders considered it feasible to re-establish a settlement on the island. Nevertheless, available beaches were more exposed than before the eruption and now consisted of stone and not sand. Plans were made for the *Puma*'s crew to improve the access to the new Garden Gate Beach to enable heavy longboat loads to be carried by bullock cart from the beach to the village.

This would be achieved by widening a rocky ravine which provided the only extremely limited access from the beach to the village on the plateau above. Soon the mission turned to this task.

Scotty had advised Capt. Mellis that the sea bottom off the Settlement was smooth and free from dangers so that the ship could go alongside the new lava if necessary. However, as the *Puma* came in closer to shore the following day, it struck what was thought to be a new pinnacle of lava thrown up during the eruption. All three blades of the port propeller were bent over, and the remainder of the voyage was completed using the single starboard prop. Later, when examined in Cape Town, seventeen holes in the ship's bottom were also discovered, fortunately none of them damaging the tanks containing fuel nor causing any further serious damage. This shallow underwater rock was marked by a buoy as a danger to shipping and is still known as 'Puma Rock'.

Landing crew and equipment on Garden Gate Beach proved to be a difficult operation and Lt Cdr Crosbie, who oversaw the shore party, thought that anywhere else in the world it would not have been attempted in these conditions as waves during the operation were well over a metre in height. The canvas-skinned island dinghy, in which the first five crew members came ashore, was swamped with seawater on landing and could be dragged up the beach only with great difficulty owing to the weight of seawater. Crosbie recommended digging out a gully in the shingle to reduce the slope, and laying wooden planks to provide a temporary slipway, with boards below the surf line weighed down with stones to prevent floating. The party got soaked, and their saturated clothing and the intermittent rain made their task challenging. A life raft was used in addition to dinghies, but it was battered by the swell, its canopy being flattened by the waves, though the crew felt more secure aboard it than in the exposed dinghies. A rope jackstay was also rigged from ship to shore as an experiment. It proved useful for transporting heavier stores and demolition gear.

The shore party's main task was to create a roadway by further enlarging a gutter which ran down the 20-metre cliff and linked the Settlement plateau with Garden Gate Beach. This was achieved by setting explosives to blast out and widen the track formerly known as Garden Gate Road, and in commemoration of that achievement the roadway has been called Puma Road ever since. Lt Cdr Crosbie and his men stayed ashore overnight. He reported that relations with the six islanders could not have been better; they worked side by side, ate the same meals, and shared sleeping space in the Residency. He thought the islanders' boat-work was impeccable, and they for their own part expressed their gratitude for the Navy's efforts on their behalf. That evening, the Falk-Rønne and Juhl were taken out to HMS *Puma* with their luggage, to await the arrival the following morning of the Dutch-owned Royal Interocean Lines ship, the *Boissevain*, for their passage to Rio de Janeiro as part of their journey home after a fortnight ashore. The *Puma* itself departed at 7.30pm that following night.

The survey party clicks up a gear

As might be expected, Allan Crawford was giving his closest attention to these many developments, and on 3rd October 1962 he wrote to John Kisch at the Colonial Office, proposing an ambitious plan involving five separate groups over the following six months:

An Advance Party carried out by a naval frigate in November 1962 to take out up to 10 tonnes of basic foodstuffs, as well as the wives and dependents of the six shore-based party plus four additional men. This would give a labour force of twelve, supported by women for housekeeping tasks, who could make the necessary arrangements in preparation for the next phase.

A Technical Services Party, carried by a naval frigate or South African vessel in December 1962, to re-open the radio station, overhaul the diesel generators and tractors, and repair the damaged water supply. This party would include one diesel mechanic, two radio technicians, twelve islanders who have worked on water and sewage systems together with up to a further 10 tonnes of foodstuffs.

Operation Tristan Return in January 1963 to carry by aircraft carrier, troop ship or cargo vessel the main body of 150-200 men, women and children who wished to return, plus their personal belongings and furniture. Accompanying them should be a doctor or nurse, a clergyman and an Administrator, should it be Colonial Office policy to appoint one and assuming that he was not already on the island. There should be 50 tonnes of foodstuffs, together with sheep dogs and livestock on this voyage, which would come from the UK, though the vessel may pick up further stores in Cape Town.

Operation Follow-On in February-March 1963, carried out by the *RSA* or a naval frigate, bringing with it a schoolteacher, agriculturalist, South African meteorologist, and any remaining islanders. It should also carry 20 tonnes of foodstuffs, plus other stores for the canteen and stamps to re-open the island Post Office.

Crawford added that the fishing vessels were likely to be at Tristan in March 1963 and could conduct a survey on the possibility of building a new factory ashore. Whether he really thought this plan was practical (possibly involving four separate journeys by naval frigates and an aircraft carrier) is hard to assess. The bold proposal must have been regarded as premature at best by UK officials who had organised support for the Resettlement Survey Party and were now anxious to receive Stableford's report on the state of the island before committing to a general resettlement.

On the evening of 8th October HMS *Puma* arrived in Cape Town. In an interview for the *Cape Times* Gerry Stableford said:

There is no physical reason why the former inhabitants of Tristan da Cunha should not return. The grass is green, the cattle are fat and .... the potatoes are edible. The useable acreage of the island has been slightly diminished by volcanic debris, but there is still plenty of room for a settlement, and a new landing beach has been conveniently provided by the winter gales. Whether the islanders want to return, and whether they should be encouraged or even allowed to return, are decisions that have to be made on political and social grounds.

Cartoon from the *Cape Times*, October 1962

An editorial comment in the same paper appeared below this splendid cartoon.

There are some people who would regard British official permission to the people to return as treason to civilisation; for the closed urban mind cannot understand anyone wanting to return to what is in effect a subsistence economy. But if the Tristan people return they will be accompanied by the applause (much of it tacit) of a vast number of people throughout the sophisticated world who would go with them if they had not been enslaved by clocks and wages and pensions and medical-aid schemes. We do not see how the British Government can thwart the islanders' desire to return; it is their general freely expressed and definite will.

In the three weeks that followed, the members of the Survey Party found their time occupied with intense activity. They hunted for mollies and their eggs; they planted potatoes and sought out those wretched dogs; they repaired the roof thatching on the village's houses; and they painted their longboats. And then on the afternoon of 1st November affairs of an international nature touched their isolated lives. At about 4pm that day the men heard something that sounded like the running of an engine. Walking out to Hottentot Gulch they could clearly see a Russian factory whaling ship off Inaccessible Island and one of its catchers off Hottentot Point. The men heard a series of bangs as harpoons were fired at southern right whales which were being slaughtered just offshore from where they were standing. Disturbingly for them the activity of the Russian catchers continued for a further two days. The Tristan population of this endangered species was greatly reduced, if not eliminated, by these unlawful actions.

Back in England, on 5th November Gerry Stableford submitted an account of his findings to the Colonial Office. Crucially he reported on the possibility of the resettlement of Tristan. Here is the essence of what he said:

> *Beaches*
> The new Garden Gate Beach was adequate for offloading smaller loads by dinghy, but with no sand or a winch it was impossible to bring ashore heavy freight using longboats or other vessels. New beaches in the centre of the lava field and east of the lava could be used if roadways were constructed from the village.
> *Buildings*
> Most existing expat houses were beyond repair and, if Government officials were to return, new houses would be required. Administrative facilities and the Post Office would entail extensive repairs and new equipment. The hospital would need a complete overhaul, combined with an audit and replacement of its equipment and supplies. The Island Store would have to be re-stocked.
> *Return of people*
> Their arrival must be in the summer months of December to April when the weather and seas are calmer and more consecutive days of good conditions can be expected so as to ensure that ships can be offloaded quickly and safely.
> *General*
> If no full-scale resettlement is allowed, the fishing company and the South African Government may be interested in the possibility of a caretaker force of families who could fish, look after fishing company property, and be available for Weather Station work.
> *Summary*
> Taking all these factors into consideration I am of the opinion that the whole community should not return to Tristan unless the following conditions are met:
> - A working party would be essential to build access roads to the beaches and improve landing facilities; to build three houses for Government officers; to

re-stock the island with stores and equipment; and to bring the island's potato growing to full production.

- An assurance would be required from the fishing company that a freezing plant would be established on Tristan and work guaranteed for the islanders.
- A minimum administration staff of an Administrator, an agricultural and public works officer, doctor and teacher must be provided.
- A wireless station would have to be re-opened either by the South African Government or the Tristan Administration.

A crisp footnote appeared from the party of Tristan islanders below these remarks:

> Island is suitable for people to return. Houses are in good condition. Beaches are as near to the Settlement as Little Beach was. There are some small beaches off what used to be Little Beach. Also, one beach at what we call 'Far Corner'. All is well. Only looking forward to seeing our people on Tristan as soon as can be arranged. Beaches are all suitable for working and unloading stores.

And so, as the southern spring progressed into summer, the men settled down to their routine of work. Bullocks needed collecting to bring the longboats over to above Garden Gate Beach; the *Frances Repetto* with Benny and Albert aboard returned from fishing at Gough with a sizeable catch of lobster, while south-easterly gales buffeted the village, making it necessary to effect repairs to one of the recently restored cottages. And in the background, there remained the ominous presence of those foreign whale-catchers. On 19th November, a Russian tanker was seen lying off Big Point with the factory ship which had been around Tristan since the beginning of the month. The following morning a helicopter from the factory ship brought a party of three Russian men ashore at 9am. They shook hands with the island team and sat down in the Administrator's house where they explained that the tanker was re-fuelling the factory ship and told them about their various whaling operations. It is fascinating to look back on this encounter during the Cold War when the islanders, as they had with all similar visitors, simply welcomed the Russian whalers ashore and entered into cheerful conversation with them. Unbeknown to them, back in Britain events were taking a new turn.

The decision to return is made

Tuesday 13th November 1962 was a decisive day in terms of the Colonial Office's approach to the Tristans and their future as Gordon Whitefield and John Kisch travelled down from London to address a meeting of the islanders at Calshot. About seventy of the men were present and so were reporters from the *Daily Express* and *Daily Herald*. The spokesmen were Joe and Willie Repetto who declared that they all wanted to go back by Christmas, or at any rate during the southern summer which ends about March. Kisch told the islanders that ministers would decide on this request shortly. Meanwhile, he asked them to give serious consideration to the points from Gerry Stableford's report. After going through the document, the islanders replied that they were confident of being able to beach the boats and get everyone ashore; large stores would not be needed. They also said that, whether a factory was re-established or not, they were unwilling to stay in England any longer. They could manage without such a facility just as they had done until as recently as fifteen years ago.

Whitefield had been told privately by several islanders that they did not want to return at the present time, but if the rest went back, they would probably follow later. Kisch informed the meeting that it would be very wrong of people who wanted to return to put pressure on those who did not. The Government's representatives had therefore decided to organise a secret ballot, possibly posing three options: to return as soon as possible, to stay in England for the time being, or to state a wish to return but be unwilling to make such a commitment now. The Colonial Office conclusion was that the request to return should be acceded to in principle and that an advance party should be mounted as soon as possible, financed by the Tristan Administration Fund which now stood at £12,000, and perhaps supported with help from Sir Irving Gane's Fund. This party would be followed at the end of 1963 or the beginning of 1964 by a general resettlement of those who wanted to go back together with a teacher, doctor and parson. Even though the islanders had said that they were willing to return whether or not SAIDC re-established a factory, the Colonial Office still wanted an administration to be re-instated. The cost of this government function had previously been covered by the revenue drawn from the sale of postage stamps and in the long run it was likely that this would be adequate to cover such an expense. Summarising the decision to facilitate the return of the islanders, Kisch later stated, 'We do not see how any other course could be made acceptable to the islanders or commend itself to public opinion or parliament.'

On 27th November Nigel Fisher, a junior minister for the Colonies, delivered a statement in the House of Commons broadly outlining the result of the meeting in Calshot earlier in the month. Here is the most telling section of his remarks:

It has been decided to help those who wish to go back to do so, but we must first arrange for an advance party of about a dozen men to make the reception arrangements.... I hope that this party will go out in a month or two's time. The others will go back towards the end of next year (during the Tristan summer) when reception arrangements have been completed and the weather will be suitable for the main operation to take place.

Two days later an editorial in the *Daily Telegraph* commented that Mr Fisher was faced with a difficulty. Not all the islanders wished to return. Most of them were only too anxious to do so – indeed their experience of life in an overpopulated, highly industrialised, noisy and complicated society had induced them to look back with yearning to their quiet pastoral existence. But such was the spirit of communal loyalty prevailing among them that the dissidents were unwilling to reveal themselves. It might indeed be necessary for the Colonial Office to take a secret ballot in order to detect those who preferred to remain here. But for his part Chief Willie Repetto was adamant that almost all were raring to go back.

On Sunday 2nd December representatives from the Colonial Office arrived unannounced in Tristan Close to conduct their secret ballot. All islanders present over the age of 21 were asked to cast a vote stating whether or not they wanted to return to the island. The voting took place in the community hall among crates and boxes, some of them already nailed and labelled 'Tristan da Cunha, South Atlantic Ocean'. This ballot was extraordinary by any democratic standards. The fact that it was unannounced meant those adults living elsewhere could not vote and no such exercise had ever been organised in England on a Sunday. The exercise was treated with disdain by the islanders, one remarking, 'There's no need for them to vote. They's nearly all done packed now, ready to go.' The vote in favour of going back was 148, with five voting to remain. However, the islanders' expectation for a full return before the end of the southern summer in March 1963 was to be thwarted by a delay in the departure of the Advance Party and the postponement of the main Resettlement Party until after the 1963 southern winter.

That very same day on Tristan brought with it a distinct sense of achievement. The morning dawned fine and after the usual Sunday service the team gathered together their guns, determined to kill the last wild dogs that continued to threaten farm animals and sea birds. At Dick's Hill the dogs were spotted in a pack at the Hardie Pond. The animals were observed by Second Watron, and by Shirt Taylor's Hut they were seen chasing donkeys, before suddenly running towards Hardie Bank and going down on the beach by the Hardie Cave. Here was the team's chance to finally out-manoeuvre the feral beasts. The team split, with Neville, Philip and Lars going by the Hardie Cave whilst Thomas, Johnny and Walter went onto the beach by the gate at the Molly Gulch to get ahead of the dogs. The ambush worked and, on that day, the very last of them was killed. Stock and wildlife were now protected, and the men could look forward to training new imported working dogs to assist with their farming jobs in the future.

The following evening the team picked up a BBC World Service radio bulletin which confirmed that fifty islanders would be leaving England the following February, and the others by the end of 1963. Whilst this was good news, they were left saddened by the prospect of another year before all the community would be re-united. The *Tristania* returned from Gough on 7th December, landing the six ship-based fishermen with their suitcases at Hottentot Beach before the vessel prepared to leave for Cape Town a few days later. The returning men would stay on Tristan until fishing resumed when the ship returned in the new year. With the shore party now twelve in number, the increased energy could be channelled into preparing for resettlement.

Interviewed by the *Daily Mirror* Willie Repetto told the reporter, 'You can't blame my people wanting to go home. On the island we were good longboat men, farmers, fishermen and cattlemen – all those things. These jack-of-all-trades talents are no good here. We are classified as unskilled.' Writing in the same paper the columnist Cassandra remarked, 'By 148 votes to 5 at a secret ballot held at their ex-RAF camp at Calshot they have weighed us and our way of life in the balance and found us wanting. Not only wanting but spiritually broke. The decision of the Tristan da Cunhans is the most eloquent and contemptuous rebuff that our smug and deviously contrived society could have received.'

In South Africa Allan Crawford kept up his campaign to accelerate plans for a complete resettlement before the end of the 1962/63 Tristan summer. He drafted a letter in December 1962 which Chief Willie Repetto wrote out and sent directly to the Queen, whom he had met the previous March. In it he welcomed

'with joy the decision of Your Majesty's Government to help us return to our Island Home. Great kindness has been shown to us by the peoples of this land, and for these we shall for ever be grateful. We are sorry that this is not our way of life. The climate does not suit us, some are sick, and others are out of work. We therefore long for our homes and the independence of our old way of life.'

He went on to say how heart-broken the people were because of the decision to delay the return of the majority until the end of 1963. Willie argued that the beaches were good for landing, especially between January and March, and a summer return would enable roadwork and preparation of ground for the 1963 potato crop to be carried out. The letter was acknowledged and passed on to the Colonial Secretary.

Christmas was now approaching. In England, the Tristan pupils at Hardley Secondary School were looking forward to the celebrations. To their credit they had made quite an impression; not only were the staff struck by their exceptionally polite and well-mannered temperament, but their new uniforms, ruddy complexions and dark hair had enabled them to stand out in their various classes. Down on Tristan, and now with the added asset of a full-strength longboat crew, eight men sailed to Stony Beach for meat. Here, a wild cattle herd is owned by the islanders, whose practice it is to kill the animals by shooting with rifles and carry out the basic tasks of butchering on-site, a tradition that continues into the 21st century. The team boiled some cuts from their kill and enjoyed a kettle of tea before travelling back with fresh meat for the festive season.

Joseph Glass at a frost-bound Calshot in January 1963 with his new dog and looking forward to enjoying the following year a warmer January. Islanders were aware they would need a new stock of dogs to replace those that had to be destroyed on the island. (Tristan Photo Portfolio)

Frank Glass begins his return home outside his Calshot home as he prepares to travel to Tilbury to board RMS *Amazon* as a member of the Advance Party. (Tristan Photo Portfolio)

Domestic cats remained in the village, helping to keep down the rats and mice on which they had thrived when the Settlement had been deserted. Aunty Martha's cat 'was heavy' when the party had arrived, and Lars now took on the task of caring for the kittens. After lunch on 23rd December a ship was seen offshore, and so it was decided to risk going out to the vessel to request it to take away their mail. Lars pushed the dinghy out at the stern but it was drawn into the sea water, whereupon a heavy swell drove the boat back on the beach. He was pulled under and the dinghy rode on his back. Luckily, he came out the other side, with little more than painful bruising to his leg. The dinghy made it out to the awaiting ship which proved to be MS *Cedarbank* and its crew took on mail. The dinghy returned and landed on the new small, sheltered beach in the middle of the lava flow. The weather proved wet on Christmas Day which, after a church service, was enjoyed indoors with a festive lunch. Several of the men who were recently married started building new houses. One of these was Lars, who rose at five o'clock one morning to yoke in two bullocks to carry sand and gravel to make 'bricks' (breeze blocks) for his house wall.

As 1963 approached Gordon Whitefield visited Calshot to confirm the list of islanders who would join Chief Willie Repetto in the Advance Party, since Willie delegated this task to the Colonial Office. A priority was to include wives of men already on Tristan, plus others with the skills needed for essential re-building and repairs. Hampshire, like the rest of the country, was suffering from the worst winter weather for many years. Mary Swain wrote that ill people were unable to get to the doctor's surgery and men could not get to work as roads were blocked by snow and no buses were running. The snowdrifts in Tristan Close were three metres deep and had to be dug through, but in next to no time blizzards had re-filled the voids just created. Tristan children had looked forward to snow, but the extreme conditions meant that the heating system at Hardley School was inadequate, and its pupils needed to wear overcoats in class. Precious little writing could be done during those bitter few weeks since their hands were too cold to hold their pens successfully.

In a reply to Allan Crawford on 3rd January, John Kisch made it clear that a final decision had been reached: to return the islanders in two stages, with an Advance Party leaving as soon as possible, and the remainder of the community following it later in the year. He added that this approach was seen as imposing no great hardship on the islanders, the press had generally accepted the decision to be right and reasonable, and that many islanders felt the same.

A fortnight later the *Tristania* arrived back from Cape Town bringing mail and stores. These included a stock of wood ordered for housebuilding, which was brought into the village by bullock cart. Letters were avidly read over a bluefish hash supper. In an incredibly significant development, the six fishermen learned that, in an act of generosity by the company, they were no longer required to go aboard for the remainder of the season. All twelve party members could therefore concentrate on preparing the village for resettlement as manpower was now doubled.

The Advance Party leaves for Tristan: resettlement is now a reality

The Colonial Office faced the problem of securing suitable shipping to transport the two parties to Tristan along with their considerable cargo. Conventional passenger and cargo ships could be used for the smaller Advance Party, but many were surprised that use was not made of the Union Castle line ships to Cape Town which departed from Southampton, which was near to where the islanders were living. Instead, it was decided to book berths on RMS *Amazon* from Tilbury to

Rio de Janeiro, with onward passage to Tristan aboard the Royal Interocean Lines' SS *Boissevain.*

Finding a ship for the main party was more difficult owing to the many challenging requirements. These included transporting some 200 Tristans in reasonable comfort, as well as conveying 200 tonnes of supplies and cargo including a couple of motor vehicles and carrying a significant number of dogs and chickens. A ship with sufficient capability for completing this 20,000-kilometre round trip was not available in Great Britain; so, interest was moved abroad. In early 1963 the Colonial Office approached the Danish *Bornholm* Island ferry company and during the next month a contract was signed to charter their flagship vessel, the *Bornholm*, for £30,000 to return the main party and their cargo to Tristan later in the year. MV *Bornholm* was less than two years old, and, in addition to Baltic Sea ferry duties, it was built as a cruise ship, carrying passengers to such destinations as London, the Channel Islands and the Canaries. The 5000-tonne, 98-metre-long vessel was ideal for the purpose and would be available during the northern winter when it was not needed as a ferry to *Bornholm* Island.

To keep costs down, the islanders were committed to help with cleaning and to work in the galley. Plans were made to build chicken runs and doghouses on the bow and to serve three daily meals to the passengers. Once news of the *Bornholm*'s charter to this far-flung destination was made public in Denmark, the shipping company was inundated with requests from people who wanted to sign up as crewmembers on a journey which caught the imagination of the public. Letters from sailors, pursers, medical personnel, and journalists flooded in, all clamouring to join the trip, but none was required as there were enough experienced permanent staff available.

On Tristan a start had been made to construct a road over the 1961 lava flow to the new small central beach. On 16th February 1963 HMS *Puma* returned and the frigate's captain sent some of his men ashore to help flatten Puma Road. The following day some of the party went out again to the *Puma* to deliver mail they had hastily written and to buy trousers, shoes, and shirts from the ship's stores. The first potatoes were now ready for digging, and the harvest yielded 7 bushels (about 200 kilos) of 'large' for food and 1½ bushels for seed. Within a month production had risen to 23 bushels (about 650 kilos) of large and 16½ bushels for seed, thereby reassuring the island team that they were generating an adequate food supply for the returning community.

On 20th February, the team vacated the Residency which had been their home since their arrival in September, and they moved into their own homes where they would now be based in preparation for the arrival of the Advance Party. The following day Albert's spirits were lifted when he heard that his fiancée Rose had picked a record especially to be played for him on the BBC radio programme *Listener's Choice.* On 25th February cattle were milked for the first time, as the opportunity for them to feed on lusher summer grass meant that they were now in a greatly improved condition. Another tradition was renewed a week later when five men took a dinghy to Sandy Point to pick apples.

Back in Britain plans were made to buy supplies to re-stock the Tristan canteen using £4000 that Sir Irving Gane made available from the Tristan da Cunha Fund. On Sunday 17th March, the Advance Party of 51 islanders led by Willie Repetto made their departure by coach from Calshot to Tilbury to board the *Amazon* for their journey to Rio de Janeiro via Las Palmas. Accompanying them would be 30 tonnes of baggage including three island longboats. Also boarding the *Amazon* was a group of twelve expatriates, led by the newly appointed Administrator Peter Day, an ideal choice for taking on this position since he had previously served in that role between 1959 and 1961. He was joined by Medical Officer Dr Gooch, Agriculture Officer Gerry Stableford together with their wives and five young children, plus a wireless operator named Morrison. Day explained to reporters that only 60 tonnes or so of light stores and luggage would be taken to the island as the beaches had yet to be prepared for landing heavy equipment. The main task facing him and Stableford was to establish the 'fabric of life' on the island in preparation for the return of the main Resettlement Party. Willie Repetto, when interviewed before boarding, explained that it was 'not possible to say anything bad about our time here, but it has been strange all the time. People have been good to us, but that has not made us happy. We know we can only be happy on our island. We shall not leave again. We have nothing against Britain, but we have seen enough snow to last a lifetime.'

On Tristan after church that morning Lars Repetto listened intently to the BBC World Service, and at 9.15 he heard the news that the Advance Party had departed, so a bottle of whisky was drunk by the team in celebration. On 28th March, while Martin, Johnnie and Lars were on their way back on a bullock cart filled with wood from the cliffs above the Patches Plain, Scotty fired a shot from the *Tristania* which was a planned signal for Lars to come on air. The great news was that on the 25th his wife Trina had given birth to a baby daughter who was going to be called Deborah. A party ensued, culminating in what Lars called a 'Braisflaise'. Tristan is still famous for its 'Braais' where fresh meat is grilled; they are always known by this Afrikaner term and never called a barbecue.

Islanders taking a meal aboard MV *Boissevain* during the passage between Rio de Janeiro and Tristan in April 1963. (Tom Fransen)

The Advance Party, 63 members strong, was soon approaching Rio de Janeiro where RMS *Amazon* docked on 2nd April, as did the *Boissevain* the following day. In Cape Town as far back as 25th February tonnes of food supplies had been loaded on the ship, destined to remain on board until 9th April, the day when it would eventually arrive at Tristan. The total cost of equipping and transporting the Advance Party's journey south was about £50,000, met from the island Administration Fund. On 4th April, the whole island-bound party transferred to the Dutch liner, by which time they had been joined by three journalists: Rhona Churchill from the *Daily Mail*, Carl Mydans from the magazine *Life*, and James Blair on behalf of the *National Geographic*.

By 8th April final preparations were being made on the island for the Advance Party's imminent arrival. The team had already taken the longboat *Canton* to the *Tristania,* just in case they needed to land the islanders away from the village if the sea was rough. The landing place was levelled by filling it in with stones, the houses were given a final tidy, and one of the other longboats was placed on the bank on top of Puma Road. Finally, at 11pm the lights of the *Boissevain* were seen approaching the island. Inevitably emotions were running high on the morning of Tuesday 9th April as the men of the island-team approached the ship by longboat to be greeted by the sight of everybody waving. They raised three cheers in response and climbed aboard to welcome them home. There was great excitement as couples, relations and friends were re-united after so many months and many wept unashamedly as is the custom even in ordinary times as family members come home safely from thousands of miles away. Happily, only a light SE wind was blowing which meant that the Advance Party could be offloaded promptly. Rhona Churchill reported the miracle of a calm sea at dawn after previous rough weather: 'There's God's hand in that,' she said, and all the men replied, 'Aye, that's God's work.'

Offloading MV *Boissevain* at Tristan onto two longboats tied up alongside on 9th April 1963. (Tom Fransen)

Churchill managed to join the first longboat ashore, with Old Frank Glass helping her and other women down the tricky rope ladder and taking an oar to row towards the beach. As the frail canvas longboat with sixteen men, women and children aboard drew near the shore, she stared up in complete amazement at the approaching 4-metre-high wall of grey lava boulders, some of which appeared to be close to a metre across. From her point of view landing looked like it was going to an impossible feat, but Frank, by now a man of 58, assumed his authority over the younger men, giving the necessary orders as he and five other rowers manoeuvred the craft closer to the beach. In next to no time thirty seamen from the two fishing vessels had tied a rope to the bow, and with a series of 'Heave-ho's' they had hauled the boat up onto the land, safe and dry.

The new beach was dwarfed by the fresh black lava just a few metres to the east. Women and girls shuddered as they stared at the cause of their evacuation eighteen months previously, and for the first time gazed at the result, the destruction of two landing beaches and the fish processing factory. While heavier stores would come ashore later, the new arrivals shouldered packing cases and strode up Puma Road to reclaim their abandoned houses, whilst a lively young collie snapped at their heels. Peter Day arrived clutching a battered suitcase, containing the precious postal handstamp inscribed 'Tristan da Cunha' and a supply of new postage stamps. No longer priced in the South African Rand, these were the contemporary St Helena shillings-and-pence issue specially overprinted with the expression 'Tristan da Cunha Resettlement 1963'. The lucrative engagement with the world's philatelic trade, on which the community would survive for income until the shore-based fishery was established, was just about ready to re-start.

Soon there was a longboat shuttle to offload passengers and luggage, all transported by bullock cart, tractor and trailer up Puma Road and on to the Prince Philip Hall which was used as a central base. Except for most of the wood, all the consignment of stores was safely brought ashore from the ship thanks to assistance from the boats and crews of the *Tristania* and *Frances Repetto*. This demanding operation was spread over twelve hours up until 7pm, when, with most of its cargo of timber still on board, the *Boissevain* set sail for Cape Town.

After so much preparation, a proper community was reforming once again. Women were reunited with their husbands, one of them being Victoria Glass who told Rhona Churchill, 'This is my happiest day ever. I's fine.' Churchill reported that she could see and hear little paraffin lamps burning in twenty cottages, pots of tea brewing, tales being told, and God thanked for everyone's good fortune. Tristan was reoccupied. Amidst all this confused activity Churchill's luggage was lost, but as she cheerfully put it, 'I's fine too'.

The following morning the two fishing vessels set sail for Cape Town at what was the end of the 1962/63 fishing season. For the now-privately owned fishing company SAIDC, it had been a great achievement. Even though there was no processing plant on the island and shore-based fishing was not yet feasible, there had been a record lobster catch of 308 tonnes from the *Tristania* and *Frances Repetto* boosted by the efforts of that group of six Tristan men fishing between September and December 1962. This success would be a fillip to prospects for a future investment in a shore-based factory.

Administrator Peter Day soon assessed the Settlement's infrastructure, a fascinating evaluation not least because it was significantly different from the reassurances that London had earlier received:

Access – Beaches were unsuitable for offloading heavy equipment except on the calmest of days, and landings were hazardous to boats and crew at all times. This ruled out shore-based fishing until proper landing facilities could be organised. The improved Puma Road was passable by bullock carts, but too steep for a tractor and trailer. A path had been constructed to the new central beach in the lava flow though suitable for pedestrian use only.

Village – Minor repairs had been made to some houses, but the majority needed renovation. Government housing and other buildings were mainly water-tight, but in a filthy state and in need of repair. The water and sewage system were working but in need of maintenance.

He also reported that seventy patches had been cultivated to supply a limited ration of potatoes for newly enlarged community of 75. In addition, about 70 bushels (2200 kilos) of potatoes had been reserved for planting in the spring.

On 12th April Tristan's Post Office re-opened for business and special first day covers were prepared, many of them signed by Willie Repetto above the title 'Chief Willie Repetto MBE', a reference to the award he had been granted in the New Year Honours List of 1959. Three days later the *RSA* arrived and, despite the swelly conditions, Albert Glass and Benny Green took a black dinghy and set out to pick up mail. On launching, the dinghy went under a swell and half-filled with sea water. When the next swell came, the two men jumped overboard, and the dinghy was still floating upright. So, Joseph Glass took off his clothes, ran in and swam to the dinghy. He then proceeded to bail her out, and row the dinghy to the *RSA*. On return he landed on the small beach in the middle of the lava field. This incident was witnessed and photographed by the press representatives present, and consequently Joseph's heroism became a subject of admiration across the world.

On 16th April Peter Day set about organising the 36 men into three gangs to work in rotation on two projects. The first, lasting twelve days, was to build a rough tractor track road over the lava field to the central small beach. The second, which took seventeen days, was an attempt to prepare a channel on Garden Gate Beach by removing gravel and large stones and shoring it up with poles. A shortage of materials and incursion by the sea caused the channel to collapse and so regrettably the project failed. For the building of such roads £1000 had been allocated by the Colonial Development and Welfare Scheme, and this sizeable sum had been brought ashore by Day who oversaw all the financial transactions in the first instance. The islanders were paid 1s (5p) per hour for this gang-work, which had been their rate of pay prior to the evacuation in 1961 but was significantly lower than what the men had earned in England. Inevitably there was a certain amount of tension amongst the men over this situation which, perhaps for the first time, set up an attitude of antipathy towards the 'government'. Indeed, some refused on occasion to take part in any such paid activities. Peter Day held general meetings when needed, and consulted with the three gang leaders, but since the Island Council was not re-established until the following year, and as Chief Willie Repetto had no special role in the new administration, Day had to take on the role of 'benign autocrat' under these circumstances.

Breaking and moving volcanic stone to make a new road across the 1961 lava field to access the new small beach that could be used as a landing place. Lars Repetto poses in the centre with his sledgehammer raised, Leonard Glass wearing a sailor's cap stands to the right, while Joe Glass points into the distance. (Tristan Photo Portfolio)

The three journalists had intended to stay eleven days, but their planned departure on 21st April aboard the steamship *Straat Mozambique* was cancelled as bad weather prevented their going aboard. Consequently, the visitors were marooned ashore. The trio had to wait until 2nd May when American naval frigate USS *Spiegel Grove* called at Tristan. Two helicopters brought ashore mail, stores, and Admiral J Tyree, US Commander-in-Chief South Atlantic, who toured the Settlement before taking on the three journalists at the end of an extended twenty-three day stay.

All the while general work continued to prepare the village for full resettlement, though not without mishap. On 4th May, for instance, the longboat *Lorna* and a dinghy took a party to Sandy Point for apples. On their return, conditions became swelly and at Garden Gate Beach the boat turned and filled with seawater to such a depth that it needed to have holes cut in the canvas to let the water out. To make matters worse, further damage was caused as the boat was drawn up the stony beach. On the other hand, the maintenance and re-decoration of the government quarters carried on apace as did the refurbishment of the Island Store, and the re-roofing and repairs to the office block. Since the *Boissevain* had left before offloading the consignment of wood, it was decided to demolish one of the wooden wartime station buildings to provide timber for repairs. Numerous modifications were made to the water system, and the inlet dam was re-built. Puma Road to Garden Gate Beach was realigned and widened for use by the tractor and its trailer. The fencing of cliff tops, the station area and the Patches was undertaken, and the road across the lava to the small beach was surfaced with gravel and soil. A generator was installed to provide lighting for the Prince Philip Hall and power for the wireless station.

On 24th May the public bar at the east end of the hall opened for the first time since 9th October 1961 and it continued to serve drinks for three hours each weekday evening. After issuing a warning about the use of strong liquor, Peter Day bought a free drink for all the men. A few of the younger islanders were thought to be drinking to excess and proving to be an embarrassment to other islanders, and so steps were taken to limit the import of spirits, thereby freeing up more shipping space for vital food supplies. Two young islanders were found guilty of theft and were each fined the sizeable sum of £50. In the words of the Administrator, who also acted as Magistrate, it was thought that this crime was the result of mixing with 'the lowest strata of English society'. Later, a by-law extended the magisterial powers of the Administrator to permit the imposing of a maximum fine of £100 as a precautionary measure to deal with any increase in crime. This early conviction for theft did much to re-establish the crime-free situation that had prevailed during Peter Day's first term of office.

A crucial shortage of potatoes was experienced at this time and this led to the need to rely on expensive imports. Only one patch per family had been planted using the seed potatoes saved from the crop dug in January 1962. A further 7 tonnes of eating potatoes had been brought aboard the *Boissevain*, but these had soon sold out as many were needed for seed. Their yield, however, proved to be extremely poor since they were intended to be eaten. To compensate for this shortfall the islanders bought more flour and sugar, which resulted in these items having to be rationed. The middle of September found the *Tristania* inadequate for transporting enough essential cargo because it was now additionally required to carry a significant volume of potatoes for the island. The vessel arrived on January 13th, but the sea was too rough to offload its contents, and it was only late in the following afternoon that a crew of men were able to take out a canvas dinghy to collect the nine bags of mail. It proved to be a fated venture. As the men came into land, the dinghy swung broadside on, and they were all thrown into the sea. Ches Lavarello struck his head, but soon recovered. Anthony Rogers, however, suffered a serious leg injury and had to be carried to hospital to be treated for a fracture which would affect his mobility for the rest of his life.

Also, on board the *Tristania* were two specialists from South Africa, the engineer Peter van der Merwe and Henry Moffatt, Consultant to South African Railways and Harbours. They carried out a survey to choose a site and plan the design of a new harbour which would not only provide good landing facilities but also be the key to re-instating an onshore fishing factory. It is hard to overestimate how greatly the future prosperity of the island would depend on such a facility.

The return of the Tristanians

Back in Britain final preparations were underway for the return of most of the islanders to their former home. By tradition, every Tristan mother was a member of the Mothers' Union. Representatives from the Winchester diocese of the organisation had greeted them on arrival and had gone on to support the community throughout their stay, and now they were preparing to bid them farewell as they returned in 1963. To commemorate their stay in Britain, a banner was made for the Tristan women to take back to St Mary's Church. On the front was a figure of 'Our Lady and the Holy Child' and on the back the words 'Tristan da Cunha Mothers' Union'. In mid-May and in anticipation of their departure this banner was presented to Head Woman Martha Rogers and two young married women bearers, and was blessed by the Right Rev. Kenneth Lamplugh, Bishop of Southampton in a ceremony at Winchester Cathedral.

As the day of their leaving drew closer, the Fawley-based Rev. Brewster wrote a 'Farewell to the Islanders' for the October 1963 issue of the *New Forest Magazine* in which he reflected on their approach to life:

> [They] have experienced our mechanical civilisation, and in many ways have enjoyed it ... but when it comes to the fundamental things which make life happy, they find these sadly lacking in a society which makes a god of money and things that money can buy. So, they are going back to their simple family and community life on their remote volcanic island, taking with them many aids to comfortable living which they have not had before, but convinced that these things do not come first. They welcome them when they are available, but they put first their families and their home, and their Faith. They remind us of the truth of our Lord's words in the Sermon on the Mount: You cannot serve God and Money. May they have a good journey and find again the peace and happiness which they value.

They were, however, assured of continued spiritual support. Rev. Keith Flint had been appointed by the SPG to accompany the islanders back to Tristan and he was now living at Calshot.

Diverging feelings, however, were being aired in public much more fully than before. On 16th October, Victor Rogers, who had voted to remain in England with his wife Lily and their children Joan (23), Rosie (7) and Roderick (2), was interviewed by the *Daily Express* and explained that the other Tristans were angry with them and they were called fools. He added, 'We are thinking of the children. In England there is a future for them. On Tristan there is nothing.'

Another notable islander who decided to remain was Adam Swain, one of the members of the Royal Society Expedition, who came back to Calshot in April 1962 and announced that Tristan was fit to resettle. Ironically, given that he was so keen in 1962 to return home, Adam decided to carry on living in England and not go back himself. He first made a career working aboard Cunard and Union Castle ships before settling permanently in Hampshire, where he was employed at Marchwood Power Station for twenty years. Altogether, twenty islanders did not return to Tristan in 1963. They included five women who were already overseas in 1961, and five who had plans to marry in England. Only ten of those that stayed would have been eligible to vote on 2nd December 1962 as the others were under 21.

Three Tristan young women met their future husbands in England and were to remain in the UK. Left to right: Mike Brown, Violet Repetto, Johnny King, Jennifer Rogers, Brian Cardy, Nola Swain. (Violet Brown)

37-year-old Louis Swain had decided to return, but did so with great sadness, telling the paper that he would be leaving a piece of his heart as he had never known such kindness. He went on to say, 'I arrived with only the shirt and trousers I was wearing. I leave with 26 suitcases of clothing and gifts.'

Back on Tristan preparations continued as the day approached when the remainder of their community would be coming home. On 8th October, the *RSA* arrived and, with difficulty because of a heavy swell, landed a pontoon to help offload the *Bornholm* the following month. Passengers returning to Cape Town included the two harbour specialists and Anthony Rogers who required South African hospital treatment for his injured leg.

Back in Tristan Close on 21st October all the islanders' belongings had now been packed and they were surrounded by just a few essentials to cover their last few days in Calshot. Until their departure, they were fed communally in one of the big former RAF dining rooms, their meals prepared by the Civil Defence Welfare Section. The following day the Bishop of Winchester, Dr Falkner Allison, assisted by Rev. Noel Brewster and the islanders' new chaplain Keith Flint, held a farewell service at the former RAF chapel. There followed a party arranged by the New Forest WVS. In an interview in the *Southern Evening Echo* later in the week the local WVS chair, Marjorie

McDonald JP, summed up the islanders as 'very genuine, unsophisticated, honest and trustworthy.... They are a delightful people to know and are very loyal in their friendships – and never gossip.' The WVS looked after the islanders so well that on their return to Tristan, Head Woman Martha Rogers wrote to McDonald to say, 'You have shared our joys, and you have shared our sorrows. You have been dear friends to us all.'

On 22nd October, the *Bornholm* left Copenhagen. Aboard were Capt. Ove Johansen, his crew and a solitary passenger. He was the writer and artist Roland Svensson from Swedish National Radio. Roland had already met the islanders in Calshot and would go on to develop a life-long friendship with the Tristan da Cunha community. Before departure from Denmark many special preparations had been made for the unusual voyage. Safety measures had been tightened, with two extra lifeboats added in the bow and a Norwegian motorboat carried to assist in the unloading of people, cargo and supplies once the ship had arrived at Tristan. The *Bornholm* sailed through the Kiel Canal and early in the morning of 24th October it berthed in Southampton alongside what was then the world's largest ship, the RMS *Queen Mary.*

The 200 tonnes of cargo, which included two longboats carefully maintained by their owners at Calshot, was soon loaded. At 4pm the islanders arrived in a convoy of coaches and ran the gauntlet of the press photographers' flash bulbs and the bright TV camera lights as they boarded the ship and were shown to their cabins. The only moment of panic arose as it was feared there was no stowage space for a large consignment of beer, but luckily shrewd crew members found room in empty cabins and so drinks could be enjoyed on the journey and back on Tristan. A further valuable cargo was a batch of over 2000 letters to be stamped in the island's Post Office during unloading and brought back on the *Bornholm* as collectable philatelic covers and attractive souvenirs of the resettlement voyage.

Joining the 198 islanders was the Colonial Office's Gordon Whitefield, who had led the liaison efforts with the community whilst it was in the UK and who had a key role in organising the voyage. He was Tristan Desk Officer in 1961 and on this trip, he was making his first visit to the island. Whitefield would return in 1965 as its Administrator, cementing his role as one of the most influential characters of this period. Also travelling with them were Radio Officer Bill Evans, the teacher Jim Flint, and coincidentally the SPG priest Keith Flint. All the passengers assembled in the ship's restaurant for a farewell speech by the Colonial Office official Nigel Fisher in which he said:

I know you are longing to be home. You have not much enjoyed our climate, our traffic, television and telephones. But whatever your views about leaving, the British are sorry to see you go. We have sympathised with you in your exile, and we admire and understand your determination to return home. You have made a great impression on the British people. I hope you won't forget us, and I promise we shall not forget you. The people of Tristan and of Britain are closer in thought and heart than before the volcano.

So, at 10.50pm on 24th October, the main resettlement voyage got underway, as the *Bornholm* inched away from its moorings and the islanders began their long journey south to their former home. The ship had been designed to cater for 300 passengers in comfort, and its modern facilities were more luxurious and up-to-date than those of the evacuation ships *Tjisadane* and *Stirling Castle* two years before. Father Flint led services on Sundays at 6.30am and dances were held in the evenings. The ship docked at Freetown in Sierra Leone on 31st October to take on fuel and fresh water after completing more than half of the 10,000-kilometre voyage from Southampton.

On 3rd November, the *Bornholm* crossed the Equator and with great shoals of flying fish accompanying the vessel a traditional line-crossing ceremony was held, during which many 'enjoyed' an early bath in the ship's pool, a ritual summed up by Roland Svensson with the words, 'many unworthy were cleansed internally and externally of the vices and the dirt of the Northern Hemisphere...... and everyone is now worthy for the joys and pleasures of the Southern Hemisphere.'

Bornholm off Tristan (Frits E. Pedersen)

Back Home at Last!

On Sunday 10th November, the *Bornholm* arrived at daybreak on Tristan, anchoring at 6.30am. Two island longboats pushed off from the shore and were soon alongside the ship, their crews rapidly climbing on board to make an emotional greeting with their families once again. The women embraced, kissed and wept silently. 'Welcome brother, welcome sister' was the often-heard cry, followed, according to Roland Svensson, by embarrassed silence, calm and solemn happiness. The main deck was packed with boxes, sacks, suitcases, and patient islanders who were led down a gangway onto a platform. Women and children with luggage went first, the swell lapping the platform and wetting long skirts before they were carefully placed on board the waiting longboats. Doreen Gooch, the wife of the doctor, recalled the arrival of the main party:

> As the old and young were brought ashore, I cried freely with all the islanders.... I greeted old Granny Jane Lavarello, aged 84, and Blind William Rogers, who died just a few weeks later, and Aunty Martha, Head Woman, whose husband [Arthur] had died rather tragically just before [his planned departure] for Tristan. The old ones just sat on the rocks and wept and said, 'Thank God, now we are back.'

Once people were safely ashore, attention switched to offloading the cargo. The Tristan men had four longboats in use with teams also aboard the *Bornholm* and on Garden Gate Beach. The Danish sailors worked calmly and efficiently at the winches

Drawing a longboat up Garden Gate Beach during the offloading of *Bornholm*, the scene as shown in a painting, *Longboat – Tristan da Cunha* by Roland Svensson. (Torbjörn Svensson)

and in the motor launch. This vessel carried far more than any of the longboats and was used to transfer loads to the kelp zone, where the launch was moored temporarily by tying with a rope to a tough kelp stem. Here loads were transferred to island longboats which were used to run the freight into the beach, where a wooden ramp made a smoother landing than on the lava pebbles that make up Garden Gate Beach. In each case a rope painter was fastened to the longboat's bow stem and a team of twenty men hauled the craft up the beach and started offloading onto higher, drier ground. Watching the beach scene for several hours, Roland Svensson became aware that practically all the island men were needed to land a boat and he concluded that the future looked an unpromising one until such time as a harbour was built.

It was expected that the *Bornholm* would take four days at the most to offload its cargo. By late morning on the 11th, some 40 tonnes (that is to say, 20 per cent of the total cargo) had been brought ashore, but the wind veered from SW to NW and became too strong for safe boat work, and so the vessel sought the lee. On the following day, the weather continued to prevent offloading, though this gave a chance for loads to be brought up Puma Road to the village. The Prince Philip Hall became the clearing area for the islanders' furniture and effects, until owners claimed their property as they settled back into their homes.

Apart, that is, from Dennis and Ada Green. On inspecting their damaged home where the thatched roof had been destroyed by a lava bomb, they decided to return to Hampshire with their daughter Valerie. Dennis had originally planned to build another house at the west end of the village, and the Tristan da Cunha Fund had provided a grant of £300 for materials but, seeing their old home ruined, the couple changed their minds and permission was given for them to return on the *Bornholm*. That sum of £300 was added to a grant from the Tristan da Cunha Fund to re-stock the Island Store, as it would benefit everybody. The remaining part of the Fund was used to purchase oars and canvas for the longboats, as well as potato and vegetable seeds. It was gratifying to its organiser, Sir Irving Gane, that the islanders had been able to take back the furniture, carpets, and equipment with which they had been supplied from the Fund in Calshot.

Altogether, £19,618 had been generously donated to the Tristan da Cunha Fund by organisations and individuals in the UK and overseas, often through local fund-raising campaigns. It is fascinating to note a sample selection from the many who gave. These included: £20 from students at a school in the United States; £17 raised by a concert held in a Northamptonshire hotel; £125 as a result of a BBC programme in

Busy scene at Garden Gate beach during the offloading of MV *Bornholm* (Jim Flint)

Islanders and their luggage on Garden Gate Beach during offloading of *Bornholm* (Jim Flint)

which islanders had taken part; £4 5s (£4.25) by Girl Guide carol singers in Hindon, Wiltshire; and £125 from the National Association of Boys' Clubs. Another significant fund-raising event was the 1962 *Daily Express* Boat Show where visitors gave £772 to the fund as they watched Tristan men constructing a new longboat using donated raw materials and tools.

The *Bornholm* was able to resume offloading on the 14th. Observing the operation, the new teacher Jim Flint watched about fifty men working on the rocky beach, hauling up longboats and towing drums ashore, each one of them containing 200 litres of paraffin. While they were floating in the sea there was no problem because they were securely roped together, but as they came ashore, all hands were needed to haul them up the steep incline. The tractor, driven by Louis Swain, carried loads up to the Settlement. On one occasion, Louis' foot slipped, the tractor ran out of control and crashed into a heap of packing cases, one of them bursting open to reveal, to the amazement of the onlookers, a coal scuttle still filled with its original contents!

The offloading of the *Bornholm* was yet again postponed on 16th November on account of the poor sea conditions, and these continued for a further three days. On the 18th Capt. Johansen contacted his company to decide how long the vessel could remain at the island as no foreign vessel had ever stayed at the island for this length of time. It was agreed that 21st November must be the latest date of departure, otherwise the ship would risk running out of fuel before reaching Freetown.

On 19th November, a strong wind was still blowing from the north-west, but offloading could wait no longer and so Capt. Johansen brought his ship to just off the Rookery, which was as close to land as he dared. From here the crew succeeded in unloading onto the beach near Jew's Point the bulk of the remaining cargo, primarily cement, iron and other building materials. The operation was dangerous and not without drama. In the words of Roland Svensson, 'Unloading under such conditions is mortally dangerous. The Danish sailors' unselfish work and the risks of working on deck or in the motorboat was a great example of excellent seamanship.' Later that day Tristan longboats transferred loads to Garden Gate Beach. Thankfully, the following day the wind backed to SW and so offloading at Garden Gate Beach could resume and continued from dawn until dusk. Furniture continued to be landed and some families were anxious that their belongings were still not ashore, especially when successive boats brought in case after case of beer and cola. People joked that there was enough drink on the *Bornholm* to float the ship or, according to Gilbert Lavarello, 'enough tins to build a new harbour'.

Offloading was finally completed late in the morning of 21st November, and all the 200 tonnes of cargo had been brought safely ashore. The scope of the consignment had been immense: furniture, carpets, clothes, cement, barrels of oil, prefabricated buildings, metal girders, a large winch, timber, hospital equipment; provisions of seed potatoes, foodstuffs, and canned drinks. There were enough groceries and other provisions to feed 264 people until crops were ready to be harvested and the next vessel paid a visit carrying with it more supplies. The *Bornholm* was ready for its long voyage back to Europe and it departed at 2.15pm, with returning passengers including Dennis, Ada and Valerie Green, Gordon Whitefield, Gerry Stableford and his wife, the radio operator Tom Morrison and Roland Svensson. As the ship pulled away, Svensson was unsure whether to say 'all is well' since he was keenly aware of the great anxieties amongst the islanders: Could the fish factory be re-established? Could a decent harbour so essential for their continued comfort and security be built? Is it ever possible to tame nature when it is capable of being so hostile?

Back in Denmark the voyage of the *Bornholm* was followed with considerable interest and its crucial role in this massive logistical exercise was viewed with great national pride. Many friendships were forged during the trip and the ship's First Stewardess said of the islanders, 'They are the nicest passengers ever to have travelled on the *Bornholm*. And this is meant as a great compliment since we always have nice passengers when running our regular service.' Without doubt the Colonial Office made a good choice when it chartered that excellent vessel, one with enough capacity, a capable crew, and the spare time essential for coping with the delays sustained so dramatically in the South Atlantic by such bad weather. For all such setbacks the *Bornholm*'s voyage had been a huge success, and the people of Tristan da Cunha could rejoice in the fact that at long last they were back home once more.

Part 6: Reconstruction and Prosperity

Getting the community functioning again

On 9th December 1963 further stores were offloaded from the Tristania to which they had been transferred before the *Bornholm* left, so there was a flurry of activity as supplies were hauled up Garden Gate Beach. The good weather also allowed shore-based fishing as boats went out for line-caught fish for home consumption. Many bluefish were scaled and gutted on the beach before being slung over donkeys which carried them up Puma Road to the village.

The Tristan school in former Naval Station buildings (Jim Flint)

Headmaster Jim Flint prepared for the start of school and discovered poignant reminders of the rapid evacuation. Pre-volcano work was pinned to the walls, a blackboard had the date 9th October 1961 marked on it in chalk, and a desk contained the rock-hard remains of a half-eaten sandwich. The school re-opened for a short term with fifty pupils on 28th November until breaking up for Christmas on 20th December. Three pupil teachers were employed and allocated to the three younger classes, with

Flint in charge of the seniors. For the five children who arrived in April it was their first formal education for more than eight months.

After problems with seed potatoes, it was hoped the 1963/64 crop would be sufficient to provide enough seed for the following year and enable the community to be self-supporting in this staple food crop. The cattle were in good shape and milk yield was very good, but only a few fences had been repaired and the overgrazing of pastures in the future remained an issue as sheep and poultry stocks were restored. New dogs would be needed to add to the seven imported during the year, with a preference given for border collies to establish a working breed on the island again, now that all the wild dogs had been destroyed. No poultry survived the evacuation, but a small breeding stock had been imported in April. The Gough team presented a few hens in May, and a further shipment of 160 six-week hens and six cockerels arrived aboard the *Bornholm* in November. In December, the stock of poultry was given out with typical fair shares: two hens per family, one each for widows and bachelors. On 13th December about thirty men set out for Stony Beach in a longboat, a barge and two dinghies to kill cattle to provide beef for Christmas. The wild herd had severely overgrazed the sparse pasture and needed thinning out. As usual, animals were shot with rifles and the party later returned with a mixed cargo of meat, fish and driftwood, which was soon loaded onto waiting ox carts or donkeys to be taken to homes in the village.

The Post Office savings scheme was re-introduced after the main party's arrival. Islanders with accounts in the UK had their balances transferred and the British Post Office Savings Bank provided specially printed bank books and forms for use on the island. By 31st December there were 186 separate accounts with over £14,000 invested although islanders were soon drawing on their savings to buy food and it was feared that money might run out before a factory started to function. Soon after the main party had arrived back on the island the basics of government employment were set in place. Pay rates were increased to 1s 3d (6p) per hour for those who worked more than ten days a month, and skilled work was paid a bonus.

A second post-resettlement stamp was issued in UK on 1st October – a 1s 6d (7½p) value as part of the Commonwealth's 'Freedom from Hunger' omnibus set – and by the end of December the Tristan Post Office had sold £2,455 worth of stamps, which was the community's only earned income in 1963. The duties of the radio operator and the postmaster were combined into a new 'Superintendent of Posts and Telegraphs' role, thereby relieving the Administrator of any postal responsibilities in a new village employment structure. Since its opening in May, the pub made a profit of £39 by the end of the year, and six films were shown in the adjacent Prince Philip Hall. These were made possible by the installation of a new projector brought by the *Bornholm*, and the re-establishment of a link with the Columbia Film Library to supply feature films through the Central Office of Information.

Drawing stone using bullock carts to repair walls. (Jim Flint)

Peter Day made some thoughtful remarks as a conclusion to his 1963 report. The physical resettlement was complete, social problems had been largely overcome, and respect for law and order had been re-established without resorting to police forces and jails but rather with traditional community sanctions reasserting themselves. Day noted that, given some means of livelihood, the majority of islanders were happy to have returned, adding:

However, they are in a precarious plight as young men build their houses oblivious of their future here, taking it for granted that the British Government will sort things out for them. Many are living on savings they accumulated in the UK, but, by the winter of 1964, many of them will face the stark reality that they have returned to an island which at present has no economy.

The basic assumption on which the return was taken was that the landing beaches were reasonable, and it would be possible to re-establish shore-based fishing. This was proved to be incorrect, and without a harbour it will be impossible to sustain a community of this size (257 islanders and 16 expatriates) except at the expense of the British taxpayer.

Planning a harbour

There is no doubt that the biggest handicap that the returning islanders faced was the loss of the safe landing beaches at Little and Big Beach, now engulfed by lava. The new landing place at Garden Gate was always problematic as boats needed to be drawn up the steep stony beach which was constantly being altered by storms. Without a harbour fitted out with proper quays and cranes, capable of acting as a haven for shallow boats, a new fishing factory would be impossible to build. And without such a factory the island would inevitably revert to a mainly subsistence economy, and islanders could well then change their minds and return to the UK.

The fishing company had funded Henry Moffatt's harbour planning visit and in December 1963 Peter Day travelled to Cape Town aboard the *Tristania* to meet him and discuss his report. Some of the schemes could have cost up to £200,000, but a compromise plan for the building of a small harbour at about half that cost was agreed. The original site chosen was the east end of Garden Gate Beach and a proposal was submitted to the Crown Agents, but they were doubtful about the wisdom of building a harbour so close to the 1961 lava field, since this was quickly eroding, and stones were being moved about by every storm. Indeed, on 13th June 1964 heavy seas deposited huge boulders in the proposed area for the harbour development and therefore this site was abandoned.

Peter Day decided to create a new post of Public Works Officer, to provide leadership during harbour construction. Alan Cox was appointed, and he arrived with his wife and two children before the end of 1963. Cox was engaged in other projects until Crown Agents chief civil engineer John Hawtrey arrived on the *RSA* in October 1964 with explosives, a compressor and drilling equipment. He first considered sites at Boatharbour Bay and sea stacks called The Hardies. These were rejected as they were too exposed to the prevailing north-west winds and would have required expensive road works as they were over 3km from the village. Instead, Hawtrey was tasked with finding a location conveniently accessible to the Settlement and with a nearby site on which to build a factory. He chose, surveyed, and pegged out a place for a small harbour at Saltpot on the western end of Garden Gate Beach and work on the project duly went ahead. Funding, now estimated at £80,000, was provided by the UK Government, with work to be undertaken directly by local island labour under Cox's direction.

In the meantime, use was being made of an 8.5-metre-long barge and heavy-duty pontoons which had largely replaced traditional longboats for offloading at Garden Gate Beach. This new equipment enabled 200 tonnes of supplies, mainly for the new harbour, to be offloaded from the *RSA* in only four and a half days. Craft were driven at full speed onto a wooden ramp on the stony beach, hooked onto a wire and pulled up the beach by a noisy, smoky engine positioned nearby. There waiting for these loads were two tractors and even on occasion the Administrator's Land Rover, the first land vehicle of any real comfort ever to operate on the island.

Constructing a road to the new harbour (Jim Flint)

A road was made over the 1961 lava field to access the central sand beach and the beach at Pigbite as alternative options whenever Garden Gate Beach was unworkable. In the event both substitute landing places were abandoned because storms destroyed each of them in the months that followed. To prepare for work in 1965, construction started in November 1964 on a road from the village plateau down the 15-metre-high cliff to the new harbour site. As part of this task lava rock was blasted away and used for the breakwater cores, enabling a road to be created that was 4.6 metres wide though with a steep gradient of 1 in 10.

Economic difficulties and social progress

Since any land-based factory had yet to be constructed, commercial fishing continued to be carried out only by the two company fishing vessels. A team of 23 island men worked aboard the ships, earning £2900 in 1964, a dramatic contrast to the £12,000 in wages earned by islanders from factory and fishing work in the last full season before the volcanic eruption. Furthermore, the start of the 1963 lobster fishing season saw catches down 20 per cent, which meant that the Tristan Government received lower royalties at a particularly critical time when it had a full population to support, and major infrastructure projects to fund. In his 1964 annual report, Peter Day was concerned that high levels of lobster catches would not be sustainable, and he therefore suggested employing a marine biologist to advise on a top limit for such catches. This is the first recorded reference to marine conservation measures that would later lead to the creation of a sustainable fishery.

The 1963/64 potato crop yield was disappointing as the imported Arran Pilot seed was planted late and affected by the wind, as well as being harmed by blight from South African seed stock. Once seed was reserved for the following season, there was a shortage for food, which meant that islanders again had to resort to buying potatoes in the Island Store. So, 51 tonnes were imported for this purpose at a subsidised price. A hundred pine trees were cut from the plantation at Sandy Point, and these were used as poles to create or repair fences around pastures, thereby reducing the need for timber imports. In April 1964, a first Nightingale fatting trip since 1961 was planned but this had to be called off, as all hands were needed to offload the *RSA* and the *AES* which brought a new stock of sheep from the Falkland Islands. They were distributed equally to island families, but this only brought the total to 206 sheep on the island compared with 657 in 1960. By Christmas as the store ran out of imported cooking oil, islanders regretted the loss petrel fat and meat that would have been obtained at Nightingale.

It was no surprise that the Island Store lost £2840 in 1963/64, as islanders had less money to spend and they concentrated on buying basic foodstuffs such as potatoes, which were subsidised by the government. A new, more hygienic, and allegedly rat-proof shop opened in July 1964, the building itself a gift from Padgetts, a Yorkshire motorcycle business. Perhaps the gift was inspired by the young islanders whose photographs appeared in the UK press with the motorbikes they had bought and ridden around Calshot. The store was a large steel-framed structure on a concrete foundation, and it was the first of many modern buildings to replace the former naval station's wooden huts, though it was still called the canteen as it had been during the Second World War.

As well as wages paid to fishermen by the company, the Tristan Government paid £4207 for work that included roadbuilding and Island Store construction, as well as a range of other jobs, some of which were administrative. Islanders were also employed directly by expatriates to help run their homes, receiving about £900 in return. This income was inadequate to cover living costs, and islanders therefore withdrew £6403 from their Post Office savings accounts during 1964, though they also received £964 in UK income tax rebates as they paid tax in full on part-year earnings in 1963/64. The island treasury banked £1300 in £5 notes during the year, all brought in cash by islanders from England as no such notes were supplied from government sources to the island. Clearly, this economic situation could not continue indefinitely, and some island families without any fishing income and with limited government work were considerately worse off than they had been in England or on Tristan before 1961. Many families had no savings in their Post Office accounts.

The Prince Philip Hall resumed its place as the social hub of the village. Dances were held and the bar opened each weekday evening, providing darts and snooker facilities. Weekly events included film shows, a children's club based around Red Cross activities, and Scottish dancing in the winter months.

Confusingly, two Mr Flints worked in the island school. Headteacher Jim Flint was now in charge since education had become a government responsibility, and Father Keith Flint, paid by the SPG, taught religious education and assisted at morning assembly. He also led a weekly service for pupils in St Mary's Church, with special permission granted by Granny Aggie (Agnes Smith was still the leader of the small Roman Catholic community on Tristan) for Catholics to attend if the ceremony was of a non-denominational nature. The school operated four ten-week terms divided by three-week holidays and an extended Christmas break. The syllabus followed the English school curriculum as there was always an awareness that the resettlement of the island might not be a permanent reality. Jim Flint observed that the span of pupil ability, from some very bright youngsters who would hold their own in a typical grammar school through to the scarcely literate, was typical of the range to be found in any English village. Even though he had been warned before arrival that the children would need strict discipline, he found in practice that the majority were quiet, polite and keen to learn.

A lagoon halfway along the new lava flow provided a temporary swimming pool thermally heated by the hot rocks beneath. (Jim Flint)

Games and craftwork were the most popular subjects and for the summer of 1963/64 the games afternoon was often spent in the warm pool in the heart of the new lava field, with a convenient shallow end for beginners and a deep end for the stronger swimmers who also dived from a jagged lava platform into the inviting lagoon. The deeper water was too hot for swimming and long strings of green algae grew there. In the winter of 1964, a storm destroyed the temporary pool by filling it with huge rocks, resulting in less frequent swimming taking place in the more exposed rockpools at Runaway Beach. Tennis and other games were played on the concrete 'Quarter Deck' in front of the wooden school building. A local game of football-rounders was already well-established, and indeed still flourishes as a Tristan favourite, having the advantage of needing only a reasonable ball and makeshift bases, and suiting both boys and girls of varying ages. Craft activities included making authentic model cottages and knitting.

Health within the community was generally good, with no dysentery cases since the late 1950s when piped water and the new sewage system had been installed. Unfortunately, 'progress' meant more use was made of Hottentot Gulch where any rubbish was unceremoniously dumped, awaiting a flood for it to be washed northwards and pollute the pristine ocean for many more years to come. A sorry practice indeed that was to continue for another twenty years.

Tensions simmer while Tristan rebuilds its infrastructure

Work continued in early 1965 to prepare for the construction of a new harbour later in the year. Some men refused to work more than ten days a month and so were only paid 1s (5p) an hour, rather than the 1s 3d per hour for more regular Government employees. All were able to earn more in England but living costs then had been much higher. Men wanted to resume their Nightingale Island hunting and gathering trips – and those to collect rockhopper guano for fertiliser were due to start in January – but fears remained that those who took part might lose out on more lucrative work on shore. In response to growing tensions amongst a reluctant workforce, Public Works Officer Alan Cox introduced a more flexible work contract, with a 40-hour week spread over five days. He allowed a day off on Saturday and promised flexibility for those who took time off for Nightingale trips or other communal work, allowing them to return to employment when their tasks were completed. As a result, a two-day Nightingale guano trip was finally held from 29th January, which, as it occurred over a weekend,

The new harbour takes shape. Note that a continuous storm ridge of boulders has now been formed, blocking off the former inlet at Garden Gate which had been used so successfully during the Royal Society Expedition. (Jim Flint)

meant only one working day was lost. A second nine-day trip went ahead in February and a fatting trip in April, thereby ensuring that Tristan potatoes were once again fried in that pungent fish-smelling cooking oil derived from petrel chicks. Harbour work may have been reduced, but Tristan's unique hunting and gathering traditions were re-established and the work force was more content.

The *RSA* arrived with the heavy equipment and supplies for harbour works in April 1965, together with a new Administrator, Gordon Whitefield, seconded from the Colonial Office for a year's term to replace Peter Day with effect from 2nd May. Whitefield was well-known to islanders through the many contacts he had established with them when they were in exile in the UK, and by joining the MS *Bornholm* voyage to repatriate the main resettlement party in 1963.

Soon blasting operations began to create a harbour pool out of older lava-formed rock pools in the Saltpot area, but the aim to create a basin 2.7 metres below mean sea level was never achieved. Therefore, only small boats up to 9 metres in length with a draft of under 76cm could use the harbour. Even for these shallow vessels, care was needed to avoid exposed reefs that had been left intact. To protect the harbour pool from damaging waves, curved breakwaters were built either side of the 15-metre-wide harbour entrance. These were galvanised, PVC-covered steel gabions filled with rock, much of it blasted out when building the harbour basin and road. Construction progressed quickly under the capable leadership of Alan Cox and the harbour started to take shape. Once this essential work was underway, the fishing company invested £120,000 in building and equipping a new processing and freezing factory on the plateau immediately above the embryo harbour.

During this period of harbour and factory construction the community suffered severe deprivation. The 1964/65 potato crop was again poor and, although stocks of sheep and poultry were slowly building up and Nightingale trips for fat and bird meat had resumed, by August 1965 the shelves in the Island Store were bare of all but the most expensive items. Flour, rice, and cereals had long since sold out and some families appeared to be living on biscuits and custard, until these items also became exhausted a few weeks before the first spring supply ship arrived in October. Post Office savings accounts were virtually empty as many islanders had spent their UK reserves. Nevertheless, paid government and factory building work was available, and those able to secure the higher rate for regular or more skilled work could buy groceries in the store to supplement their scarce potatoes, limited supplies of home-grown vegetables, and cuts of meat.

On Easter Monday 1966 blasting operations shifted rocks that blocked the new harbour entrance. Dynamite was tied to sandbags which were dropped over the side of a dinghy. The detonation was effective, but it proved to be a terrifying experience even for those standing way up at the top of the cliff who found themselves showered with dirt and stones.

Building of the company-funded fishing factory continued apace as prefabricated sections were erected on its concrete base. Large cylindrical fuel tanks were placed alongside, linked to new generators. The fishing vessel *Gillian Gaggins,* (which replaced the *Frances Repetto* in 1965) brought a first load of diesel oil on 18th April and this was pumped ashore through a lengthy black heavy-duty hose stretching a kilometre out to the anchorage. Before long, the new generators were installed and produced electricity for the factory, an essential requisite for powering the new freezers, with enough capacity also to provide the Settlement with electric current in due course.

An April Nightingale fatting trip meant once again that labour was short for essential harbour and factory construction, but the prospects of stocking up with cooking oil and salted petrel meat proved too attractive. The trip was fraught with difficulties. A man was severely burned when a Primus stove exploded, several women were sick, and a first attempt to return home failed as the boats were forced by high seas to return to Nightingale. Finally, with food noticeably short and more members of the group becoming unwell, the boats returned in poor weather, got caught in a squall, and were obliged to remain at sea overnight. They were carrying emergency flares, but these did not work, and they were forced to land the following morning at Stony Beach, damaging two longboats in the process. Three men trekked over the mountain to alert the village to prepare a rescue operation, but there was a downpour that evening, and so it was only the next day that men set off on foot and by barge. Finally, the longboats made it home by late afternoon, but the seas were still heavy as the last boat was hauled up the beach. Eventually their cargo of oil, birds and stores were loaded onto bullock carts and donkeys, and the members of the hunting trip were able to return to their homes and to the warm celebrations awaiting them. Stocks of food on Tristan were practically zero by this time, but the *RSA* was fortunately at Gough Island with 100 tonnes of stores aboard and within a month commercial fishing was able to resume.

Even during construction, the new harbour was in use and, when the rebuilt fishing factory was ready, the inaugural commercial fishing day was held on 23rd May 1966. For the first time, loads of lobster were transferred by crane from the new fishing dinghies onto waiting trailers on the quayside to be pulled by tractors up the short, steep harbour road to the factory above. There, a team of processors was ready to clean, pack and freeze the precious crustaceans ready for export to the waiting world markets. Further harbour work continued, and it was officially opened and named Calshot Harbour in a ceremony on 2nd January 1967, the name being chosen as a tribute to the Hampshire village that had been home to the islanders in 1962/63. Now the dream had been realised, as Tristan's fishing economy, linked to a new harbour, was back on track. Islanders had a source of income, some full-time with the fishing company, others part-time processing or catching fish as they could be released from their government jobs on fishing days. The people of Tristan could again afford a wider range of goods in the Island Store and start to save to improve their homes. But for all that, some islanders missed out on this renaissance.

Returning to Britain has its attractions

During the early re-building phase, a significant number of islanders became increasingly discontented, and many began to form plans to go back to the UK. First to follow Dennis Green and his family were 'Big' Gordon and Susan Glass who started their journey to Southampton aboard the *RSA* in May 1965. Gordon was the islander who was the victim of the mugging attack in England in 1962, but his daughters Lilian and Ada had stayed in England and, since he had lost an arm several years before in a fish-factory accident, he considered himself too handicapped to resume a full life back on Tristan. Following them in November 1965 were Ned and Dolly Green since Ned needed medical treatment, which in those days was provided in England rather than in South Africa. Also travelling on the ship was Gerald Repetto, then aged 20, a capable young islander who worked as a teacher under head Jim Flint from 1963. Gerald was now free to plan his own future and returned to England, later joining the Australian Navy but returning to live on Tristan in 2005. On arrival back in the UK, he claimed in an interview with the press that 75 per cent of the islanders would return to England if they could, but that the old men were keeping them back. In fact, the islanders were indeed free to return to England if they wanted, and a significant group were giving thought to doing just that, buoyed up by an offer of assistance arranged by Administrator Gordon Whitefield. Whitefield was aware of the reluctance of some islanders to travel back to Tristan when working at Calshot in 1962-63 and wanted to help them return to England to fulfil their wish. A list of 39 names of people wanting to leave was drawn up in October 1965, but some changed their minds, and so it was a group of 35 islanders who, on Easter Sunday 10th April 1966, boarded SS *Boissevain*, the same ship that had brought the advance party to Tristan in 1963. A departure of islanders is always emotional, and in the pouring rain that morning there were scenes of sorrow because so many were leaving, at a time when a few others staying behind felt unsure about their own future on the island. In South Africa, the party transferred to the *Capetown Castle* and they arrived back in Southampton on 3rd May. Therefore, out of the 280 Tristan islanders who could have returned home in 1963, no fewer than 66 – that is to say, nearly 24 per cent of them – had either remained in the UK or had subsequently moved back there.

The large group of 35 returning to England that May was not all unified in their purpose. Families were now split between England and Tristan, so difficult decisions had to be made. The situation in England had changed too, as those who returned were not refugees and could no longer rely on special help from the Tristan da Cunha Fund or the Colonial Office; they had to fend for themselves. Indeed, the Colonial Office

Islanders, including Harold Green (looking right), wait to disembark MV *Capetown Castle* at Southampton Docks on 3rd May 1966. (Tristan Photo Portfolio)

itself no longer existed as in 1966 it was merged into a new Commonwealth Office, which in its turn morphed into the Foreign and Commonwealth Office (FCO) in 1968. The New Forest Rural District Council had re-let its homes in Tristan Close, but it did its best to re-locate the new arrivals who were now scattered around the district, with some of them living in caravans while they waited for proper housing to become available. Jobs were still obtainable, but the community lacked its previous cohesion and many quickly changed their minds and returned to Tristan. There was certainly no longer any public sympathy for their plight after so much help and support had been provided to enable the refugees to settle down in 1961 and to facilitate their return to their island home two years later.

Harold Green and his family lived in a small, terraced house on the main A35 road in the Totton suburb of Southampton with a back view onto a brick wall and noisy traffic at the front. He told a visitor he was 'like a wild bird in a cage' and his family stayed less than a year, even though his brother Dennis, living in Holbury 15km south, remained happily in England. Dennis advised those coming to England in 1966 to settle down and enjoy life in Hampshire as he had done, by working hard as they used to on Tristan and earning money to enjoy a good life. Harold's family was part of a group of eight who started their second journey back to Tristan aboard the RMS *Windsor Castle* on 21st July 1967, and he was destined three years later to become the territory's first elected Chief Islander. Back on the island Harold would often remind people that 'There's no place like home.'

Some other returners stayed longer, but also changed their minds and went back to Tristan. Sidney and Alice Glass's eldest daughter Valerie had remained in England. She had left Tristan in 1952 with the schoolteacher Mrs Handley, who arranged for her to be educated in the UK, later becoming a teacher herself and settling permanently in England. As a result, Sidney and Alice were drawn back to Britain, together with their daughter, Trina with her husband Lars Repetto (stalwart of the Resettlement Survey Party). When Lars and Trina returned again, they were accompanied by two children, both born in England: Debbie in 1963 and Paul in 1967, as well as Sidney and Alice.

Some of the returners to England, like Dennis and Ada Green, settled permanently. A key reason to remain in the UK was to keep families together. Arthur and Edith Repetto, with their son Martin, returned to be near their daughters Dora, Mildred, Violet and Vivienne, all of whom married English men and settled happily in Hampshire, but leaving behind Ernest and Michael who were both married and remained on Tristan. Victor and Lily Rogers continued living in Tristan Close with their three children, after choosing to stay in 1963. By 1970, 35 of those who changed their minds had returned to Tristan and 31 islanders remained in England or had moved further afield. Dora Tarrant (née Repetto) remained in Tristan Close all her life and was the last islander to live there until her passing in January 2013 aged 84. It should also be noted that throughout Tristan's history the population has fluctuated, and many islanders have emigrated to settle elsewhere, especially to the USA, South Africa and England as well as Germany and Australia.

A modern village grows and prospers

In April 1966 Gordon Whitefield was replaced as Administrator by Brian Watkins who oversaw further development following the opening of the new harbour and factory. Under his guidance paved roadways replaced the rough tracks in the village, but it was unfortunate that he chose the names of administrators, including his own, to identify the different roads without consulting the islanders who no doubt shrugged their shoulders, knowing such a proposal was not welcome and would most likely be ignored. So, there is no sign saying 'Watkins Road' in the modern village.

Following the successful installation of diesel-powered generators at the new factory, electricity was installed in the village. Streetlights were erected, so Watkins is also remembered as the Administrator who lit up the Settlement. After government buildings were connected, island homes were themselves fitted with electric lights and, in due course, electric-powered kitchen appliances such as refrigerators (known as coolers on Tristan) and radios. Unless it was a fishing day, electricity supply ceased at 10pm, so candles were always handy, as were torches for walking late at night around the village when there was no full moon.

In Watkins' frank memoirs he admitted struggling with finance and losing £5000 in his first year of overseeing the Island Store. So, he recruited an expat treasurer to ensure that the growing government economy functioned more efficiently. Watkins was also shocked by an assault when an islander struck another with a stone. This led him to appoint as Tristan's first policeman the respected young islander Albert Glass, who received training at the Hendon Police College in England during 1967. Returning as Sergeant Glass he was supported by two other reliable young men as special constables to maintain a watch on public order, especially at the bar alongside the Prince Philip Hall. As one islander put it, 'Albert is all right; he don't cause any trouble.' Tristan islanders prefer to deal with any misdemeanours quietly, and such matters usually remain confidential within a community where all are related to one another. The imposing presence of Sergeant Glass patrolling the village became a familiar sight for over twenty years. It is a mark of the respect in which he was held that Albert was twice elected Chief Islander in between the three terms served by Harold Green. Between them the two men occupied the post until 1988.

Labour relations had been difficult before and after 1961 as islanders wanted higher pay, and also asserted their right to take time off for various activities such as tending animals and crops, working on their houses, and undertaking trips to Nightingale. Against the background of those previous conflicts of interest, the island's fishermen formed a committee in July 1968 and demanded £1 for each basket of 45kg crayfish caught. This was turned down by the company, and the men went on strike. They consequently lost pay, though they could continue their farming activities, their private fishing by dinghy, and selling some meat and vegetables to expatriates for cash. Also, for part-time fishermen, government jobs continued. The situation deteriorated when there were rumours that the company might try to break the strike by bringing in fishing vessels to fish Tristan waters, and threats started to be made. Administrator Watkins feared an outbreak of violence and despatched, for the only time in his long government service, a 'Flash' telegram asking for a Royal Navy ship to be sent to the island. The frigate HMS *Naiad* was off South America and was ordered to steam to Tristan. Meantime the Administrator was asked by both sides to arbitrate. A compromise was reached with a graded pay scale of up to 15s (75p) per basket, and happily the only confrontation in which the *Naiad*'s crew were involved was a football match against islanders.

After his two terms as Administrator both before and after the volcanic eruption, Peter Day had moved on to work for the fishing where he became Managing Director. This meant that he represented the firm during the 1969 dispute, gaining a bad press from several writers as a result. This has the effect that Day, like perhaps every Administrator, is more remembered for his less popular decisions. Yet his achievement in master-minding the preparation for the return of the main settlement party in 1963, and the determination he showed to ensure a harbour and factory were built, mark his tenure in the post as pivotal in securing the success of Tristan's revival. The labour policy under which he rewarded only those prepared to take up regular work on essential infrastructure projects (such as new roads, an improved landing place, and later the harbour) was controversial, but ultimately effective. In the context of his senior managerial role with the company in Cape Town, Day's influence on the island's affairs continued for another thirty years. This later position allowed him to monitor fishing on Tristan and to liaise with the Tristan Government and former colleagues in the new FCO to considerable effect.

In 1969 Day wrote a report in which he highlighted how Calshot Harbour was never built to the specifications laid down before his second term as Administrator ended in May 1965. He drew attention to the inadequate harbour basin, planned to be 2.7 metres deep and now recommended work to at least achieve a practicable depth of 1.5 metres, which would allow its use at all states of tide. He also suggested the use of dolosse to protect the quickly corroding gabions from further damage from wave erosion. Dolosse (singular 'dolos', so called after the Afrikaans name for a sheep's ankle bone, the shape of which resembled that of the concrete system) are tetrapod concrete blocks then only recently invented and already in use in South Africa. Day arranged an engineer to visit Tristan, and as a result a dolos mould was made up in Cape Town. The concrete dolosse were subsequently constructed on a site east of Calshot Harbour under the direction of George Hannah, an energetic Scot, and this work was carried out in 1971.

Calshot Harbour in about 1976 showing the interlocking dolosse installed in 1971 to reduce wave damage. (Nigel Humphries)

Chief Willie Repetto retired from his largely ceremonial post in 1970. The Tristan Island Council lacked formal status, and Peter Day had generally preferred to gather all the family leaders together whenever he wanted to consult effectively. In 1970 Island Council elections were introduced with all those over 18 eligible to vote for eight councillors. These were to be supported by three appointed members – the resident priest, the Factory Manager, and the Island Store Manager – with the Administrator as President. Harold Green became the first elected Chief Islander in a separate poll amongst the eight councillors. A new Administration Building had come into use in January 1969 which housed offices for the Administrator and Heads of Department, a handsome Island Council Chamber, a library, reading room and, on the lower floor, the Post Office.

Governor Dermod Murphy attended a council meeting in September 1970, and he was not surprised to be met mainly with silence from the newly elected councillors, except for a short speech of welcome by Harold Green, as he had experienced similar reticence on St Helena. On a tour of the factory, he was impressed by Assistant Manager Basil Lavarello, returned from work overseas, and considered him suitable to become manager in due course. The Governor inspected the new hospital under construction on higher ground at the west end of the village, following the British tradition of cottage hospital building on sites which provided good ventilation for convalescence. It was named Camogli after the home of the two Italian settlers Andrea Repetto and Gaetano Lavarello and opened the following year. A new Island Store Supermarket followed in 1972 but would still be called the canteen.

Growth was funded using fishing royalties and increasing income from postage stamps, rendered particularly popular after the widespread media coverage surrounding the volcanic eruption and evacuation. International philately became extremely lucrative in the 1970s and 1980s, buoyed by the higher disposable incomes of many collectors, effective publicity, and a regular range of attractive commemorative issues. Tristan da Cunha was in an ideal position to exploit this trend with world-wide interest in the extraordinary community that had survived such disruption and with the production of many special stamps reflecting the island's history and environment. From 1969 the new Post Office under the Council Chamber took advantage of this expanding market, and profits from the overseas sales of postage stamps began to eclipse those from Tristan lobster sales. In 1974 stamp revenue was £60,000, compared to £44,000 in royalties from fishing company profits, although islanders also earned considerable wages from their work with the company itself.

The school was the last facility to remain in the old wartime station buildings. An open site set to the east of the village was chosen for the replacement. Here was built a handsome cluster of classrooms, library and large hall with a stage set around a central quadrangle. St Mary's School opened in June 1975 with 67 pupils in five classes. Between the new school and the sea cliffs was the American Fence which provided an ideal playing field as well as a cattle pasture and heliport.

Smart and practical Colt bungalows, arriving as prefabricated sections and speedily put together, sprang up to accommodate expat staff and their families. All were built on concrete bases, some of which were the foundations of the old naval station sheltered from the wind by mature flax hedges. These bungalows were set in the middle of the village, below islanders' homes and above the burgeoning business district on the lower settlement slopes which contained the factory, offices and stores. They were much lighter and drier than island homes which generally lacked a damp course and were usually built into the hillside with no rear windows.

Administrators usually served for two years, but Brian Watkins remained for three years up to May 1969 when he was replaced by James Fleming. Fleming proved controversial. During his second term he introduced a by-law banning islanders after work hours from entering government-owned areas of the Settlement, including his Residency home. This was not acceptable in a society where all land was communally owned and there was universal freedom of movement. When he went on leave again in 1974, Rev. Jack Jewell and Acting Administrator Bill Sandham received petitions signed by island men from Chief Islander Albert Glass, requesting that Fleming should not return. The approaches were taken seriously, and there followed a visit to Tristan by the Governor, Sir Thomas Oates, in November 1974.

After interviewing eleven islanders as well as expatriate officers, the Governor concluded that Fleming should not return to Tristan as he would command little or no respect or authority from either the officers or the wider community. Therefore, determined islanders had instigated the quiet dismissal of their Administrator and in so doing sent a clear message to Whitehall that government should be by agreement with the community. Despite this closing episode Administrator Fleming had overseen a period when Tristan continued to prosper. Respect for the office was restored by Stan Trees who was Administrator from 1975 to 1978. In his last Island Council meeting as President, Chief Islander Albert Glass paid tribute saying, 'Stan has been a decent bloke.' High praise indeed!

By 1978 revenue exceeded expenditure by £146,000 with reserves of more than two years' spending. Islanders had £47,350 in their savings accounts, and they sought to enlarge their homes and equip them with a range of modern appliances and furniture, each family spending around £900 to fulfil those aims. Employees contributed 3 per cent of their salaries to a pension fund which provided a retirement income to men and women at 65 years of age and to widows with dependent children. Tristan's standard of living was improving out of all recognition.

Tristan Island Council in about 1977. Left to right - Back Row: Lewis Glass, Basil Lavarello, Lars Repetto, Benny Green, Barton Green, Stanley Swain, Ken Green; Front row: Harold Green, Chief Islander Albert Glass, Administrator Stan Trees, Father Edmund Buxton, Pamela Lavarello. Harold Green and Albert Glass were both elected Chief Islander for three terms of three years and occupied the post from 1970-1988. (Nigel Humphries)

Tristan da Cunha's viability as a community depends on shipping to transport goods and passengers as there is no airport or any realistic prospect of one. Despite the island's extreme isolation, requiring a sea passage of 2810 km which typically takes at least seven days, Tristan has adequate basic sea access for essential passenger and cargo transport. At the core of this service are the regular sailings of fishing company vessels to and from Cape Town, provided as part of a contract with the fishing company. The 1085-tonne former Icelandic ship MFV *Hekla* came into service in 1985 and was able to carry twelve passengers, unlike the smaller ship MFV *Hilary* which it replaced. The *Tristania II* replaced *Gillian Gaggins* in 1974 and continued in service until 1996 when it was replaced by the 1678 tonne MFV *Kelso.* In that same year, the *Hekla* was re-named MFV *Edinburgh.* These regular sailings were augmented by an annual visit from the South African Government's Antarctic Research Vessel, as part of a contract for South Africa to run their meteorological station on Gough Island established in 1956. The *RSA* featured in this role until it was replaced by the 6123-tonne SA *Agulhas* in 1977. The new ship provided comfortable accommodation with the probability of prompt landing by helicopter, whereas other vessels needed safe harbour conditions which could delay coming ashore by several days.

In 1985 RMS *St Helena* began annual visits, normally from St Helena to Tristan, and on to Cape Town. The ship plied from the UK, allowing British goods to be imported directly by sea to Tristan. The RMS, as it was always called, was also popular with visitors who were able to enjoy a high standard of accommodation, food and service aboard the ship. In 1990 a new RMS *St Helena* came into service and continued regular passages to Tristan until 2004, with a capacity to carry 128 passengers compared to 72 on the old vessel. The RMS service helped join up St Helena, Ascension and Tristan da Cunha and enable the Governor to complete a convenient triangular voyage from St Helena and back via Tristan and Cape Town.

An imported bakki, offloaded from SA *Agulhas,* being secured on a barge ready to bring ashore. (Nigel Humphries)

Tristan's good fortune continues

The community's mounting prosperity continued into the 1980s and this was reflected in several areas. Reserves had reached £798,000 by 1984, thanks to fishing company royalties which had risen that year to £176,000, and by stamp sales which peaked at £328,000 in1982. The latter was boosted by the lucrative Royal Wedding issue commemorating Prince Charles's marriage to Lady Diana Spencer the previous July. Personal wealth of islanders increased along with rises in wages, with government departments paying a total of £61,536 and the fishing factory £86,990 to employees in 1983/84. A good income could be obtained by energetic fishermen, paid by catch and coupled with their ordinary government or factory salaries. Island Store receipts totalled £164,910 in the same period.

A particular concern was the high cost of electricity, which was supplied from factory-based generators using expensive diesel oil imported from South Africa. Therefore, the Island Council decided to invest in renewable wind power and a Danish-built generator was installed in April 1982. It was located at Hottentot Point to benefit from its exposed position to gain the maximum benefit from the prevailing north-west winds. Councillors were provided with a projection that the wind generator would give an economic return on investment within five years by providing fuel savings of about £9000 per annum. Tristan Investments were sceptical about the £80,000 scheme and refused to back the project financially. Plagued by operating faults from the outset, and only operating well for a short time, the turbine came to grief during the night of 25th May 1983. Less than a year after it was built, it was destroyed by a severe storm, wrecked by the very winds that enticed the island to invest in this renewable energy source. The scene of destruction centred on the turbine's heavy-duty motor lying forlorn beneath the contorted turbine tower. Elsewhere, the three generator blades had blown away, one ending up in the sea and one miraculously landing just a few metres short of the factory manager's house where Basil Lavarello and his family slept. Attempts to get compensation failed and the Island Council would in future be very wary of spending precious reserves on alternative energy sources.

Expatriates and their families numbered 31 in March 1982. This group included an Administrator, Treasurer, Doctor, Agricultural Officer, Head of Post Office and Communications, Superintendent of the Public Works Department (PWD), Education Officer and Senior Teacher, a padre, seven wives and fifteen children. To help run their homes most of them employed island women, who also provided home-grown potatoes and often brought fresh fish caught by their menfolk. Pensioners frequently worked as gardeners and even though expats might have preferred to be independent, here was a way to help the local economy and be part of the community. Most posts were administered through the Overseas Development Administration who recruited staff and paid the bulk of salaries into a UK bank account as part of the UK Government's Overseas Aid Scheme. The Tristan Government provided a house at a nominal rent and paid expat officers a local salary, equivalent to similar island pay and sufficient for island living costs, which were low compared to the UK. Whilst these low costs might save expats money, there were drawbacks since they missed out on professional training and often returned home without a job through the inability to apply in time or attend for interview. A Tristan posting was therefore a risk for career-minded professionals without a permanent FCO contract.

During the 1980s it was decided to localise the leadership of several island departments, as there were capable islanders to run them, some of whom had benefitted from a more rigorous local education and overseas training. Following the earlier appointment of Sergeant Albert Glass as Head of Police and Basil Lavarello as Factory Manager, Dereck Rogers became Agricultural Officer in 1983, Lindsay Repetto and Stan Swain led the PWD and Jean Swain became the island's Treasurer. When in January 1992 Marlene Swain became Education Officer, there remained a greatly reduced 'station'

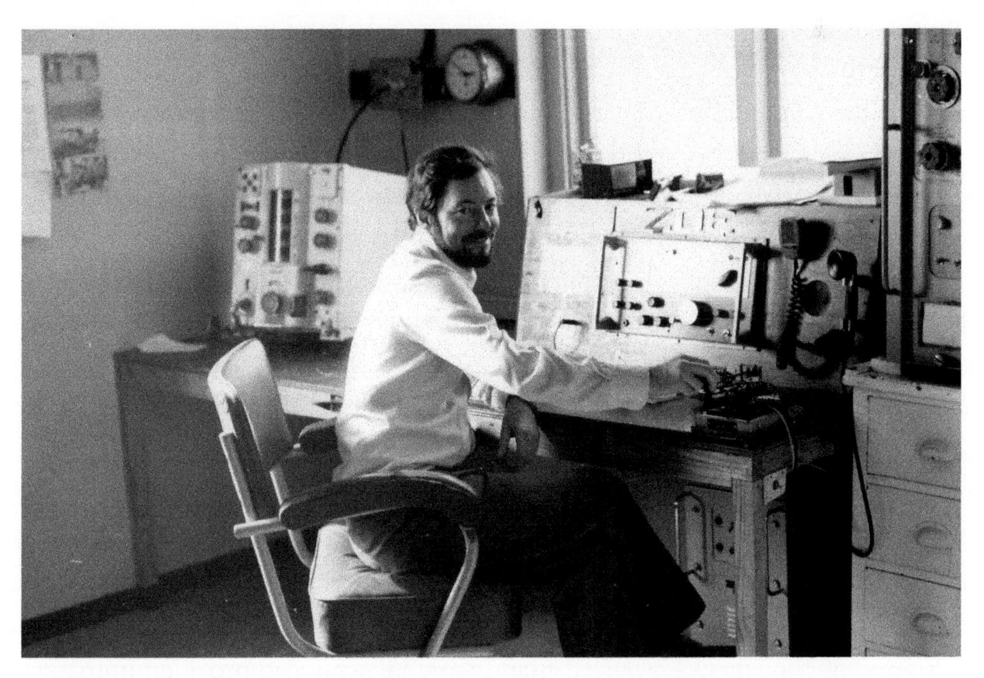

Pat Patterson at his desk in the radio building in 1976, his right hand poised to send another message by Morse code to Cape Town Radio, in cypher if it was official. (Nigel Humphries)

consisting for a time of only the Administrator and doctor and, intermittently, various priests. These were now provided by the Diocese of Cape Town rather than the SPG as had traditionally been the case.

The Head of the Post and Communications Department, a Geordie named Pat Patterson (alias Postman Pat) first came to Tristan to build a new radio station in 1976. Returning in 1982, he married Susan Green in 1983 and so became an islander, leading the Post Office during its most profitable period. Social life on the island moved with the times, and one of Pat's tasks was to organise the introduction of a new cable TV system which led to Monday night Prince Philip Hall cinema shows being replaced with the regular showing of videos on sets in island homes. Pat and Susan returned to the UK in 1984 but when Pat retired, they moved back to live on Tristan in 1987 where they stayed for ten years before deciding to settle in Hampshire. And so, Patterson was only a temporary addition to the island's well-known surnames. Allan Swain took over Pat's post in charge of post and communications, partly as he, and his deputy Andy Repetto, could use Morse code skilfully. Code was still used on Tristan in the 1980s because the audio quality of radio communication was poor and official government business was transmitted in cipher. A perk for expats – and one that gave some consolation for life in isolation – was having a Morse code message allowance of a hundred words a month to send and receive regular telegrams to families back home.

Administrators were typically FCO civil servants, often with other overseas experience. The first, Hugh Elliott, was also an eminent naturalist, having helped to set up the Serengeti National Park in what is now Tanzania before being seconded by the Colonial Office to become Tristan Administrator in 1950. Roger Perry, who served two terms between 1984 and 1989 was also not a typical FCO civil servant and had a much wider background than many of his counterparts. He was an experienced naturalist, explorer, and author, working with the BBC Natural History Unit, serving as Director of the Charles Darwin Research Station on the Galapagos Islands, and working as a Wildlife Adviser on Christmas Island before arriving on Tristan. His book *Island Days* provides a fascinating insight into 1980s Tristan life. Perry thought the existing administration was designed for a far larger community, and, whilst he accepted that there was a need for what he called 'superintendents' of departments, he was astonished to find that each had a sub-committee of the Island Council to oversee their operations as well as separate committees for the Island Store and the Prince Philip Hall. While he acceded to the former, he drew the line at a separate Finance and General Purposes Committee which he promptly disbanded as he thought it 'was one of those brainchilds of the civil service designed to draw together the haphazard ends of government'. Describing Sergeant Albert Glass's less than onerous duties he noted that the island jail was used to keep people out rather than in as it had been requisitioned by the Island Store to hold liquor securely. Perry had an early insight into the island's alcohol intake as he watched 200 cases of brandy, stacked along similar quantities of wine and sherry, together with 72,000 cans of beer being loaded onto the *Agulhas* before it left Cape Town on his arrival journey. In Perry's five years as President of the Island Council, he could not recall a single vote that was needed to make a decision as councillors normally accepted what sub-committees had proposed. Yet he seemed impressed that members always attended in their Sunday-best clothes and no doubt looked up to an Administrator such as himself with his own smart dress sense, always seen walking to work with a smart hat, polished shoes, jacket, and tie, often with a silk pocket handkerchief to match.

Perry developed a fondness for the islanders and especially enjoyed the company of pensioners. When meeting neighbour Mabel Lavarello at her half-open stable door she welcomed him with the 'I's so glad you's come', adding later that she expected Perry to get to grips with the people who dropped litter round her house. Later he asked Lizzie Hagan about the changes that had occurred since the eruption of the volcano, to which she highlighted having dry clothes, medicines to quickly relieve pain and Tristan Radio to listen to. Perry employed a pensioner to tend the Residency garden and raise the Union Jack flag daily on the adjacent pole. Gilbert Lavarello took over this role on retirement, and Perry, a renowned botanist, was highly amused when told by Gilbert one early spring 'you may call them snowdrops but we calls them white daffodils'.

Perry was impressed by Chief Islander Harold Green who was elected for a third term in 1985. Harold was one of the most outspoken of Tristan's fishermen during the 1968 strike and took a leading part in another wage dispute with the company in 1983. Harold accompanied Perry to attend talks in London which focussed on negotiating a new fishing concession contract. It was the first time an islander had taken part in such a discussion and proved the forerunner for future Chief Islanders to attend what became Ministerial Councils of the UK Overseas Territories. On this occasion, experienced fisherman Harold cut across arguments put across by the company's legal counsel with a quiet ''old yer 'orses' with a direct manner that perhaps had never been heard at Whitehall. In between sessions, Harold delivered a Tristan crocheted shawl for PM Margaret Thatcher at the door of 10 Downing Street. Thatcher was much-admired by islanders, as she was regarded as the one who went to the protection of fellow UK citizens on the Falkland Islands earlier that decade.

Despite the many modern developments brought about by a cash economy, time-honoured activities such as farming, hunting and gathering continued to flourish. As well as cattle for beef and milk and sheep for mutton and wool, many hens were kept, and some pigs were reared. Pork was a popular meat, but pigs needed a lot of expensive feed and so they were reared only for a few years. If a government sow 'done pigged', a draw was held to decide who could buy one of the piglets – another example of the equality that still pervaded the village.

Saturdays saw islanders busy at the Potato Patches where the production of this staple crop was more than adequate for local consumption and for the following year's seed. Other vegetables were grown, especially pumpkins, run in the patches after early potatoes had been dug. The Agriculture Department sold some tomatoes and salad crops grown in the Mission Garden established originally by SPG priests. It was now protected from gales, since it stood surrounded on three sides by the steep black walls of the 1961 lava flow and was expanded by the planting of more New Zealand flax hedges to provide further shelter. Here was a small silver lining to the eruption, but although this initiative showed promise, the range and quantity of fresh fruit and vegetables was extremely limited and people relied on imported goods, most frequently in frozen, canned or packet form.

Conservation and education inspire the creation of the Tristan Association

The unique flora and fauna of the Tristan da Cunha islands had long been recognised by scientists, including those aboard HMS *Challenger* in 1873, the Norwegian Scientific Expedition in 1937/38, and by the GISS in 1955/56. In 1968 Martin Holdgate, by then Deputy Director of the Nature Conservancy in the UK along with former GISS botanist Nigel Wace, returned to Gough Island via Tristan together with islanders Harold Green and Herbert Glass. Their objective was to look for changes resulting from twelve years of continuous human occupation, not least following the move of the South African Weather Station from its original site to a new building at the south end of the island in 1961. They found some unwelcome arrivals, including slugs and centipedes, almost certainly imported by GISS in 1955, and a new invasive creeping pearlwort Sagina procumbens, possibly carried from Marion Island. Nigel Wace also recorded a dead rat in a crate, but thankfully no other members of the species had succeeded in making it ashore alive. Back on Tristan they also observed how the

1961 lava was being colonised by plants and insects. Following their visit Wace and Holdgate assisted in the compilation of a new Conservation Ordinance for the Tristan islands which, with the support of the Island Council, was enacted by the Governor of St Helena in 1976. The text was included in an important monograph on Man and Nature in the Tristan da Cunha Islands which was published that same year.

The Settlement Tristan da Cunha, a painting by Roland Svensson inspired by his final visit to the island in 1983. (Torbjörn Svensson)

That Ordinance contained regulations that protected wildlife by law and helped preserve the islands' unique habitats. A nature reserve was created on Gough Island, thus paving the way for later World Heritage Status. The ordinance also protected threatened whales, seals, sea - and land-birds. Yet it permitted managed harvesting of petrels, and the gathering of petrel and rockhopper eggs, especially on Nightingale Island. Here was a pragmatic partnership between conservationists and the Tristan community that recognised the value of a responsible hunting-gathering tradition within a precious habitat.

The study of wildlife became a significant focus of interest on Tristan. St Mary's School employed a second teacher from 1977 to take charge of the older children. The first in that role, Christine Stone, established a new science curriculum and helped pioneer a nature reserve at Jew's Point to protect and study the nearest rockhopper colony to the village. A formal Certificate of Secondary Education (CSE) course in 'Tristan Islands Studies' was introduced in 1982 which enabled pupils to study local science, geography and history in depth. Pupils needed to complete a written research project and produce a traditional handicraft product. Pensioners taught model long-boat-making skills, and one handsome boat was sold by a pupil to a passenger aboard the *QEII* for a noteworthy $50. The annual trip made by the *Agulhas* was a highlight of the school year as Capt. Bill Leith, who was consistently well-disposed towards the island, allowed the vessel's helicopter to give schoolchildren aerial rides from the school's American Fence playing field.

Molly and chick on a typical pedestal nest in the Tristan Base study area. (Richard Grundy)

Members of the Denstone College expedition on Inaccessible Island's Salt Beach in 1983. Left to right: Leader Michael Swales, islander guides Nelson and Harold Green, students David Briggs and Richard Holt. (Denstone Expedition to Inaccessible)

The annual *Agulhas* trip also provided transport for scientists who carried out field work on Gough, Nightingale and Inaccessible Islands. This work was led by John Cooper and later by Peter Ryan at the Percy FitzPatrick Institute of Cape Town University, both of whom later became honorary Tristan da Cunha Conservation Officers. A nature reserve was set up by the school to study breeding mollies which nested on the Base above the Settlement. Staff and pupils hiked up the notoriously dangerous Hottentot path onto the mountain, mapping nest sites, checking eggs and monitoring chicks. Rings and pliers were supplied by Cape Town University to enable the school teams to ring both adult and mature chicks during the season, with details sent to the university to add to their data base. Agriculture Officer Dereck Rogers, who showed great interest, led some of the bird-ringing and provided boats to access Jews Point reserve. One pupil who took part was Trevor Glass, who became Head of the Conservation Department when it was founded more than twenty years later.

Michael Swales, the former GISS vertebrate zoologist, organised a successful Denstone College Expedition to Inaccessible Island in 1982/83 which produced the first detailed map of that island. Swales was Head of Science at the Staffordshire school, and as a generous gesture to the Tristan community, the College governors awarded sixth form scholarships to island students from 1983 until 1998. Four of the recipients lived overseas in 2021, but Iris Green returned to be Head of the Tristan Post Office, while Renee Green (née Glass) taught in the island school.

Swales' return to Tristan for his second island expedition also inspired him to join with Allan Crawford to create the Tristan da Cunha Association (TDCA). A foundation meeting was held in 1987 at the Royal Geographical Society in London when Crawford was elected chairman, with Swales as secretary and treasurer, and islanders Vivien Baker and Lorna Lavarello as committee members. Martin Holdgate was a founding Vice President who had served as Chief Scientist at the Department of the Environment and was at the time Director General of the World Conservation Union in Switzerland. The Association continues today to help the island, provides a UK-based forum for Tristan matters, and re-unites many of those who have lived and worked on the island. In the early days, the former Administrator Peter Wheeler arranged meetings in his New Forest home village and a bi-annual newsletter was inaugurated.

Jim Kerr taught in St Mary's School from 1985, becoming Education Officer and serving until 1992. He continued to strengthen Tristan Island studies and maintain the Base molly study which was later taken on by the islanders. In 1972 the school-leaving age in Britain had been raised to 16 but on Tristan, children were able to finish school after their 15th birthday. Keen pupils, all girls, returned for after-work examination classes but this was a poor arrangement. From 1983 an experiment enabled pupils to take CSE examinations at the age of 15, a year before their UK counterparts. Thus, both boys and girls gained CSE (and some GCE 'O Level' qualifications) a year early, one achieving success in five subjects and allowing early access to 'A Level' courses in the UK. Here was proof that islanders had the potential for higher education. Tristan raised the school leaving age to 16 more than twenty years later. However, with full island employment the continued appointment of islanders to professional jobs like teaching and nursing without the need for higher education qualifications remains an anomaly in a world where such professionals need minimum standards of education. There persists, therefore, little incentive to obtain qualified status for those content to stay on the island.

Jim returned as education adviser from 2009-2012, and again in 2014, so becoming the longest-serving expatriate to work on the island. Kerr's return heralded the re-introduction of qualified teachers to work in the school. The islanders included Anne Green who became Head on Marlene Swain's retirement in 2015.

Five island girl students travelled to South Africa or the UK for further education

St Mary's School staff and pupils assembled in 2009. Head Anne Green is on the far left, and Education Adviser Jim Kerr is far right on the back row. (Jim Kerr)

7th November 2017 was the 200th anniversary of the day that the original settlers, Samuel Burnell, William Glass and John Nankivel, signed their partnership agreement, the island's first constitution, and in celebration, a special event was held at the British Library. In this photo Tristan students (left to right) Rhyanna Swain, Janice Green and Jade Repetto, who were studying at Peter Symonds College in Winchester, supported by grants from the Tristan Education Trust Fund, are being shown the original 1817 agreement by a library curator. (Chris Carnegy)

between 2013-2019 and all were supported by grants from the Tristan Education Trust Fund organised by the TDCA. It will be interesting to see how many of them settle back on the island to use their enhanced skills to benefit the community. Low educational achievement compared to UK norms, especially amongst boys who are readily offered work without any qualifications, is a weakness that is becoming a greater handicap in the high technology-based world of the 21st century.

Part 7: Facing further challenges in a changing world

Continued threat from natural disasters

During July and August 2004, there was another volcanic eruption, heralded by tremors which shook the Settlement. Bags were packed as memories of 1961 were stirred, contingency evacuation plans were made, but the earthquakes ceased, and all was calm. In the weeks that followed, fishermen discovered and collected blocks of pumice floating around the island's southern waters. An early report indicated the epicentre of the seismic activity was 59km south-west of the Tristan Settlement beyond Inaccessible Island. This was corrected by Vicky Hards from the British Geological Survey to a location 50km SSW from the village. In 2020 a detailed new volcanic study concluded that the 2004 eruption was 43km away due south about 12km due east of Nightingale Island. Confusion indeed.

The fascinating 2020 paper identifies a seamount named Isolde which appears to be a very young volcanic cone rising from the seabed to a height of 300m below sea level with a summit crater in the order of 500m wide. No fewer than two thousand earthquakes occurred in this area between July and December 2004, and there were two more in 2012. The authors raise the possibility that the present-day hotspot plume may be centred under Isolde seamount and not under Tristan da Cunha island. It also shows that the 2004 eruption was not of the Tristan volcano, but of a separate submarine volcanic centre. It is premature to rule out any further volcanic eruptions on Tristan itself, which must still be regarded as active. Yet it does look as if the Isolde volcano is growing and could in the distant future become a new island close to Nightingale.

Much work has been undertaken since 2004, led by Dr Anna Hicks, now a committee member of the TDCA, to produce a Tristan Disaster Management Plan. Evacuation exercises were carried out in 2011 and 2014, and a Disaster Management Building was erected and partially equipped at the Patches. Dr Hicks drew attention to the weaknesses of the island's preparedness for a future volcanic eruption, one major flaw being an inadequate monitoring network which cannot precisely locate the posi-

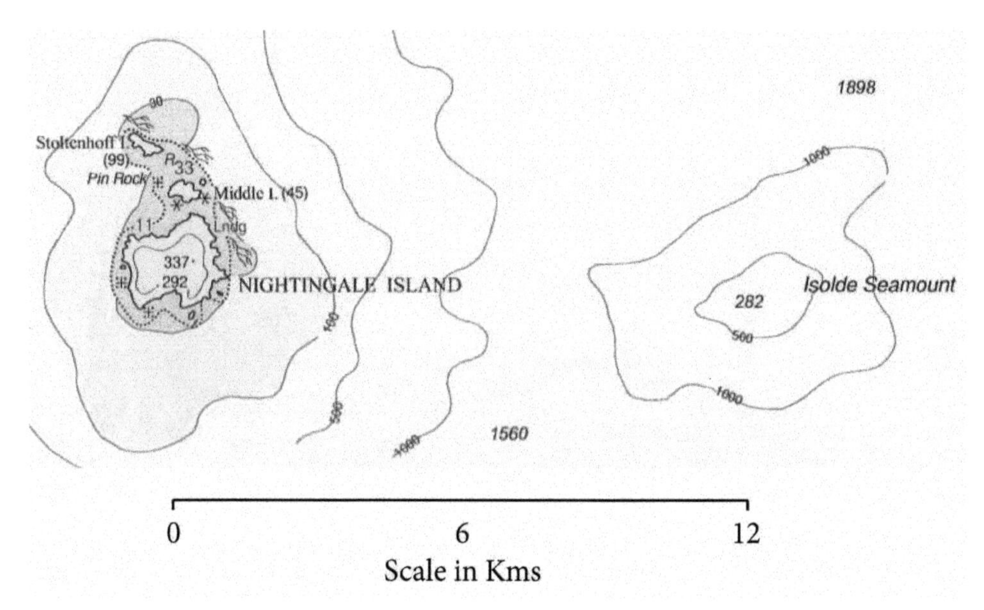

Edited 2020 Admiralty Chart extract showing the location of Isolde Seamount.

tion of earthquakes - often a crucial warning sign for an impending eruption. The problem is that the seismometers on Tristan (see the later reference to the CTBTO station) fail to provide lateral data and what is needed is the establishment of at least two more instruments, ideally located on Nightingale and Inaccessible Islands. These would offer an effective local triangulation network capable of accurately pinpointing the focus of earthquakes, and thus reducing the uncertainty of the location of possible future volcanism. A swarm of earthquakes recorded in a submarine setting *near to* Tristan poses a very different hazard and therefore necessitates a different response to a swarm of earthquakes recorded *beneath* the Tristan edifice. Dr Hicks has repeatedly raised this problem, at local and UK government level, but little action has so far been taken. In other parts of the world, with airports or shipping immediately available, evacuation of vulnerable people can be carried out relatively quickly. The Tristan community remains isolated and was truly fortunate to escape so easily in October 1961: their volcanic monitoring and preparedness to escape another emergency remain inadequate.

Extreme weather events also cause problems. A severe storm in May 2001 caused no death or injury, but it devastated the village, especially the vulnerable new buildings unprotected by flax hedges and by the massive rock-hewn gable ends of the traditional Tristan cottages, many of which had roofs torn away by the gale. The TDCA immediately launched a Disaster Fund which raised £30,000 to support reconstruction.

Damage to the factory, hospital, and the Prince Philip Hall was particularly severe and the latter was the last public building to be repaired, its re-opening being celebrated with a dance in September 2004.

But such celebrations were short-lived. Further destructive storms in July and November 2019 caused massive new damage especially to government buildings which were exposed to hurricane-force gusts. Indeed, these ferocious winds removed roofing from many of the structures on the island. These included St Mary's School, the Post Office and Tourism Centre, the Police Station, and the Administration Building. The storms also destroyed PWD stores and ironically completely destroyed the disaster management building at the Patches. Many departments occupied temporary premises during the clear-up and it took until July 2020 before staff and pupils resumed education activity at St Mary's School. Unlike in 2001, the UK Government immediately pledged funds to repair most of the damage when the first storm struck in July 2019. The TDCA also launched another Emergency Fund to help get the school back up and running, and to provide a contingency reserve for the future. It is notable that in 2001 local labour managed to repair all the damage, but in 2019 contractors were brought in as the local PWD was not able to complete the work, partly through a lack of staff.

These storms were exceptional, but floods and landslides are a regular occurrence on the Settlement Plain. Rainfall normally percolates into the porous ash soils on the Peak and is absorbed by plants and peat soils on lower slopes, with streams rising on beaches, or at a few places across the plains, such as at Sandy Point and the Big Watron stream to the west of the village. During very heavy rains the soil reaches saturation point and there is run-off into the normally dry gulches which radiate from the Peak. Streams become torrents and, when in spate, they carry massive loads of debris, including stones and large boulders. When these rock-charged streams reach the Base edge, they create massive cataract-like waterfalls down the cliffs. These in their turn smash down onto the plains below, depositing outwash fans of rocky debris which destroy pasture, block roads, and threaten buildings like the huts at the Potato Patches. Sometimes, large areas of unstable rock near the top of cliffs collapse, producing massive landslides with huge debris piles at the cliff base. Managing such damage is a routine part of living on Tristan da Cunha, but attempts are being made to monitor rainfall patterns more scientifically and to ascertain whether global climate change is likely to make violent storms more frequent. If successful, these could help predict the future mass-movement of rock debris which might once again threaten the Tristan Settlement if it occurred behind the village as did the eruption of 1961.

The 1961 lava flow has been colonised by introduced garden plants like hydrangea and arum lilies, here framed by an example of metal waste which unfortunately mars much of the area as it has been simply dumped by the side of the road and not buried as intended.
(Richard Grundy)

Roof damage to St Mary's School in July 2019 (Robin Repetto)

Fishing thrives despite emergencies on land and sea

Tristan da Cunha's economy remains centred on the continuing success of its fishing industry. The SAIDC, operating as Tristan Investments, continued to run the fishery until the concession was taken over by Premier Fishing in 1997. This event meant that Peter Day retired from his role at the fishing company, but he continued to be associated with Table Bay Marine which provided shipping services to the island until shortly before his death in 2021 at the age of 94. So, his long association with Tristan covered sixty-two years back to 1959 when he first became Administrator. His second term in 1963-65 was crucial, as was his further involvement with Tristan as managing director of the fishing company. In these roles he wielded more influence on Tristan da Cunha's affairs than any other expatriate. Further change followed and the South African company Ovenstone Agencies (Pty) Ltd acquired control of the fishing contract in 2003, remaining the concession holders to this day. Under their Managing Director Andrew James and working closely with Head of Fisheries James Glass, the company has overseen major developments to ensure the success of the key tasks of protecting, catching, processing, and marketing Tristan rock lobster.

View of the lower Settlement and Calshot Harbour in 2005. The extended 1960s fishing factory is the large building set above the harbour, with its oil storage tanks far left. The 1992 western harbour quay extension provided considerably more shelter from swells. (John Hodgkiss)

Calshot Harbour underwent an extension to its western quay in 1992 but it could still be accessed only by shallow-hulled inshore vessels in calm sea conditions which limited its use to about a hundred days a year. In 1999, the engineering consultant Robin Webb recommended the construction of a new harbour at the Garden Gate site, originally considered in 1963, at an estimated cost of £16 million. This alternative was rejected, partly because the cost had increased to £21 million by 2004.

There began a series of expensive harbour works. Breakwaters, notably the exposed western arm, were extended in 2004 to offer more protection, but this did little to aid operations. A further £2.8 million refurbishment project to the existing harbour was announced in 2006. Further damage occurred that year, and in 2007 it was recognised by the British Government that the harbour was at risk of collapse if repairs were not undertaken with immediate effect. A bold initiative saw the setting up of a UK task force, code-named Operation Zest, comprising a military team travelling out on the 16,190-tonne Royal Fleet Auxiliary *Lyme Bay* in early 2008 to carry our harbour repairs. On its way south, the *Lyme Bay* received the startling news from the island that on 13th February 2008 a devastating fire, caused by an electrical fault, had completely destroyed the cliff-top fishing factory.

The vessel arrived in perfect weather conditions on 28th February and began work strengthening and improving the exposed west harbour wall, building a concrete crest slab on top of that wall, and reinforcing the quay wall with steel and concrete. More than twenty Royal Engineers slept in the Prince Philip Hall, whilst in the surrounding area, tents provided a kitchen and dining room for the troops at a base known as 'The Cat's Whiskers' after the Panther emblem of RE 34 Squadron.

In parallel, Ovenstone mobilised help at great speed and MFV *Edinburgh* arrived on 5th March with an insurance assessor and design engineer to inspect the factory ruins. The vessel also carried company officers and a new electricity generator to replace the one destroyed in the fire. On 14th March 24-hour electricity was restored to the Settlement once again. Even though 9 tonnes of frozen lobster were destroyed, it was near the end of the fishing season, and MFV *Edinburgh* was able to catch the remainder of the Tristan quota in March. The Task Force volunteered to help Factory Manager Erik McKenzie clear the factory site on Easter Sunday, always a strictly adhered-to holiday for islanders, and Easter Monday was celebrated as Queen's Day when all the community joined in thanking the Operation Zest team for what they had achieved. When Capt. Paddy McAlpine delivered the keynote talk to the TDCA the following year, he reported that the eleven-week mission had cost more than £7 million.

Any euphoria was short-lived, however, because on 11th April the boom of Tristan's heavy-duty crane collapsed and had to be scrapped. This limited cargo lifting to 4 tonnes, so neither the large barge nor the fisheries rigid inflatable boat (RIB) *Wave Dancer* could be launched. Offloading was so slow during the next *Edinburgh* trip that fifty crates were returned unopened to Cape Town after the ship spent twenty days at the island. As a result, there was a shortage of groceries, and stocks of flour ran out. It took until April 2009 before an impressive and substantial steel-framed crane became operational behind the south quay of the harbour.

A 2017 view of the village showing Calshot Harbour and, on the plateau above it, the latest fishing factory, built to replace its predecessor, which had been destroyed in the 2008 fire. (Bernard Pronost)

Foundation work on the new fishing factory started on 1st November ahead of the arrival of the *Baltic Trader* on 19th December 2008 bringing with it the heavy-duty materials and plant. The steel frame was in place by February 2009 and in March new generators arrived together with factory fittings as the project entered its final phase, leading to its completion in June. Trials took place ahead of the start of the fishing season on

1st July, and the official opening by 93-year-old Alice Glass was celebrated on 17th July, only 520 days after the fire. So, sixty years after the opening of the old Big Beach canning factory in 1949, later to be destroyed by lava in 1961, and 43 years after the opening of the second factory, here was an up-to-date building designed to meet stringent 21st-century EU specifications and take the island's fishing industry into the modern era.

The busy scene, photographed in 2020, as a group of fishing boats return to Calshot Harbour, showing the crane, right, used to manoeuvre boats to and from the water. The larger gantry crane allows considerable loads to be lifted ashore from pontoons to enable large modern building projects to go ahead. (Kelly Green)

The impressive Operation Zest was quickly put into perspective by Administrator David Morley who described the work as 'emergency first-aid'. Gaping holes in the western breakwater were soon discovered and, starting in 2009, a rolling series of projects added more dolosse, installed a new crane, repaired quays, cleared boulders washed in during storms, and further deepened the harbour basin. Altogether an impressive total of £24.5 million was spent on the harbour between 2008-2020, a level of support which demonstrates the UK Government's continued commitment to enable the community to function. Unfortunately, Tristan da Cunha will always be battered by powerful waves, which means that harbour works will be a perennial expense. If the alternative Garden Gate scheme proposed by Robin Webb had gone ahead, it too would no doubt have also needed similar on-going maintenance and should not be regarded as some lost panacea. Tristan's sea access via a harbour will always be expensive to maintain, and so a constant drain on the UK Treasury. The reality has to be faced: no good harbour site can ever exist on a circular volcanic island lacking sheltered natural bays.

The 6400-tonne oil platform *A Turtle* marooned on the seabed off Tristan's south-east coast in 2006. (Sue Scott)

Three 21st-century incidents have disastrously shown that although modern shipping has the benefit of the latest equipment aided by satellite navigation, human error and accidents still occur, and the Tristan da Cunha islands remain vulnerable to damage from shipwrecks. The first of these events occurred in 2006 when a huge 6400-tonne oil platform *A Turtle* was being towed by the heavy-duty tug *Mighty Deliverer* from Brazil to Singapore. High seas caused the vessel to release the platform, whereupon it drifted off and – perhaps unbelievably – by 6th March it was deemed lost. It was not until 7th June that it was eventually found, when a party of Tristan islanders going to Sandy Point for beef cattle spotted the enormous structure, marooned on a reef some 250m offshore from Trypot Beach. A salvage tug hired by the owners arrived on 22nd June and there followed a saga of failed attempts before the rig was finally floated. It was soon towed out to sea where it 'turned turtle' and was scuttled at a depth of 3500m in a position some 10km east of the island, to form the islands' biggest shipwreck. Damage to the environment was limited to a small oil slick which soon dispersed, and to the arrival of a shoal of Brazilian porgy fish which had been inhabiting the rig's environs and provided an unwelcome challenge to marine biologists anxious to protect the islands' unique marine ecosystem.

The 75,300-tonne MS *Oliva* marooned on rocks at Spinners Point on Nightingale Island on 17th March 2011. The photograph shows the ship being abandoned as the Greek captain makes his way precariously down a rope ladder into a waiting RIB, which here is completely hidden by the swell. The inflatable boat had come from the cruise liner MV *Prince Albert II* which just happened to arrive that same day. (MV *Prince Albert II* crew)

The second wreck was caused by a catastrophic catalogue of errors aboard the modern 75,300-tonne MS *Oliva* during its planned passage in 2011 between Brazil and Singapore with a full cargo of raw soya beans. The problems included: failure to carry the detailed BA Chart 1769 of the Tristan islands; lack of an electronic chart display and information system; a chief mate who was considered not fit to stand a watch; and the fact that when passing within 6km of Inaccessible Island the land mass was thought to be rain clouds. The vessel's crew saw but ignored a white light later confirmed as the MFV *Edinburgh* fishing off Inaccessible, and moments after a second large echo had also been dismissed as a rain cloud the *Oliva* rammed straight into Spinner's Point on Nightingale Island early on 16th March 2011.

The *Edinburgh* was summoned to help, and Capt. Clarence October organised the transfer of twelve *Oliva* crew to the fishing vessel as a 'voluntary evacuation'. The weather worsened during the night, and a decision was made to abandon ship, but sea conditions prevented this happening using *Edinburgh*'s fishing boats. Just when more help was needed, the cruise ship MV *Prince Albert II* fortuitously arrived at Nightingale Island. RIBs, planned to be used to bring passengers ashore on Nightingale to view the wildlife, were launched and managed to steer in between the listing vessel and the rocky shore. Somehow the remaining crew clambered aboard, the last to abandon the ship being the Greek captain. True to a generous tradition that has evolved over nearly two centuries, the crew were welcomed ashore at Tristan and looked after with no arrests or recriminations. Early on 18th March the stricken ship broke in two under the force of a relentless swell. Soon afterwards 1500 tonnes of stinking fuel oil began to leak from the wreck and 65,000 tonnes of whole raw soya beans seeped into the previously pristine sea.

Having left Nightingale after their annual moult, numerous rockhoppers became covered in diesel fuel from the stricken MS *Oliva*. Many of the birds struggled ashore where they were corralled into fenced areas before being transported to Tristan for cleaning up.
(Tristan Conservation Department)

Unfortunately, the wreck coincided with the end of the moulting season for rockhoppers. Their period of starvation was complete, and they were ready to return to sea, but now that route led directly into the wretched oil slick. Two members of the Tristan Conservation Department already based on Nightingale quickly corralled as many birds as possible, providing drinking water and preventing them from entering the polluted sea. Other oil-smothered birds struggled to return to shore, now unable to swim, and made a sorry sight. An impressive operation sprang into action with resident RSPB Project Officer Katrine Herian working with Tristan Head of Conservation Trevor Glass, officers from the Southern African Foundation for the Conservation of Coastal Birds and a large team of islanders. Thousands of oiled rockhoppers were transported in special boxes to Tristan da Cunha where a rehabilitation centre was established in a large PWD building. Here birds were rehydrated, fed, cleaned up, moved to the island's swimming pool and prepared for sea release in a hugely professional operation. Nevertheless, the vast majority of rockhoppers were too weak to survive after fasting during their period moulting ashore. Tragically only 318 birds were successfully released to sea out of the 3718 rockhoppers brought to Tristan.

Tristan's crucial fishing industry was effectively put on-hold after the *Oliva* ran aground. The Nightingale and Inaccessible fishing grounds were closed as they became polluted with oil, leaving an uncaught quota of 61 tonnes of lobster in the waters around the two islands. High pressure hoses pumped warmed sea water to remove the oil from polluted rocky shores. No detergents or chemicals were deployed in the process, but booms were used to help contain and remove the oil. The Tristan Government obtained an undisclosed payment from the ship's insurers which covered all costs. These included recompense for the loss of fishing, and provision for future monitoring, which in turn considerably enhanced facilities, equipment and inshore boats for the fishing and conservation departments. The *Oliva* was soon swept into deeper water as it further broke up and disappeared from its wreck site at Spinner's Point. The remaining oil cleared naturally, and the soya beans seemed to cause no serious harm. Rockhoppers returned to the islands in subsequent seasons and before long Nightingale and Middle Islands appeared pristine once again. With that, commercial fishing for Tristan lobster could be safely resumed in the area.

In the year of the *Oliva* wreck, the Tristan da Cunha fishery was awarded Marine Stewardship Council (MSC) accreditation. It was examined against the MSC standard for sustainable fisheries and passing that assessment demonstrated that the fishery is managed in a way that does not lead to over-fishing; that it maintains the health and productivity of the wider marine ecosystem; and that it has an effective management system in place. On receiving the award, the Director of Fisheries, James Glass said:

> The careful management of the lobster fishery is the foundation of the continued economic independence of the Tristan community. MSC certification will, we hope, help us keep and find new markets, and the globally recognised eco-label will help customers differentiate our lobsters from others.

Tristan lobster cases. (Ovenstone)

By 2011 strict rules governing the size and type of lobster caught (smaller individuals and those with eggs are returned to the sea), as well as annually reviewed quotas for each of the islands, were in place. These allowed that lobster yields of over 400 tonnes can be obtained, and the industry is sustainable into the future. The main sales of raw lobster tails to American and Australian markets, and raw or cooked whole lobster to Japan, were complemented in 2014 by exports to the European Union. For that the fishery needed to meet strict European requirements for food safety. The EU accreditation of the modern Tristan Factory in August 2013 paved the way for full access to EU markets in October 2014. Andrew James, the Managing Director of Ovenstone, welcomed the news saying:

> We are delighted that the Tristan Lobster will soon start appearing on menus throughout the EU. These are the most exceptional lobsters in every sense and while we could easily supply the entire annual harvest to our current markets, it has long been our ambition to offer our product to the European market. It is no secret that

> distributors across Europe have chased us for many years to do so. Now that we are finally here, we hope the market appreciates the lobsters' unique provenance and MSC-certification alongside their superior flavour.

Soon Tristan lobster products were on the menus in top hotels and restaurants in Europe, and the delicacy was on sale in Selfridges prestigious Oxford Street food hall to attract the interest of its more discerning clientele. The result of this increasing success is the crucial income paid by Ovenstone in royalties to the Tristan Government. Together with other fishing licences, this has provided the mainstay of the island's income, exceeding £800,000 in recent years.

Tragically, a third disastrous shipwreck occurred in October 2020 when the latest Ovenstone factory fishing vessel MFV *Geo Searcher* struck a submerged rock during routine lobster-fishing off the north coast of Gough Island. The ship soon took on water and started to list badly, leading to Capt. Clarence deciding to abandon ship. Sixty crew and two Tristan islanders working aboard as fishing observers managed to clamber into life rafts and, towed by small power boats, they were able to get clear of the scene as the 1863-tonne vessel rapidly sank below the waves. The flotilla made its way southwards on an arduous sea journey, taking four hours to cover the 16km to Transvaal Bay on Gough's south-east coast. Here, staff from the South African Government Meteorological Station used the island's crane and sling to winch all 62 survivors to the warmth and safety of the base on its plateau site above the cliff. Their great escape from danger was made. SA *Agulhas II* was summoned to rescue the men from the crowded Gough Base, and all returned safely to homes on Tristan, South Africa or beyond.

MFV *Geo Searcher* listing prior to sinking after being holed when it struck an underwater rock on 15th October 2020. (SAMSA)

The outer-island fishing season ended abruptly with the loss of the ship and its crop of lobster. Fortunately, about 100 tonnes of lobster caught by the *Geo Searcher* around Inaccessible and Nightingale Island had previously been transferred to the sister ship MFV *Edinburgh*, for onward transfer to Cape Town. The *Edinburgh* (built more than fifty years ago in 1970) was quickly re-fitted for lobster fishing and travelled out to the islands in January 2021, while the *Baltic Trader* was brought back into service to replace the *Edinburgh* for regular cargo and passenger voyages.

There had been concern that Britain's decision to leave the EU might lead to tariffs on Tristan rock lobster exports, but, in an agreement that the European Union struck with the USA which came into force in December 2020, the EU eased restrictions until at least July 2025 for US lobster imports, and this was extended to other lobster imports worldwide, including those from Tristan da Cunha. So, customers in Europe's top restaurants, hotels and shops can continue to relish a great culinary taste unhampered by tariffs. It would therefore seem that this core Tristan trade remains secure into the near future at least.

Improved communications bring many benefits.

The internet has transformed Tristan da Cunha's communications with the outside world. Access to dial-up internet was available in the Administration Building from 2000 and replaced the old Morse code messages using radio. There followed an internet-linked phone service to all Tristan homes and offices, with islanders being connected as if they were living in London. Soon phones were in regular use in the village to check on friends and relations, to arrange to meet or pass on news, and to speak to relatives and friends overseas. The former Post Office became an Internet Café where islanders could access the web and use email, as they were later able to do in their own homes when the local wireless network improved. Television, relayed from satellite signals with help from the UK forces' welfare charity Services Sound and Vision Corporation, gradually improved until islanders could watch a range of live BBC and ITV programmes, and SSVC's own channels too. Together with downloaded films and DVDs, islanders were no longer cut off from the rest of the world but increasingly up to date with global affairs. This revolution in communications also put islanders culturally much closer to the UK. Since their return in 1963 socio-political influences had come more from South Africa than from Britain, through shipping links with Cape Town, stays in Government-owned and -managed Tristan House in that city, and

In October 2008 the *Agulhas* helicopter lifted a team onto the Base at Burntwood to install a new VHF Repeater Station. From the left: Ivan Green, Patrick Green, Aron Swain, Head of Communications Andy Repetto and Mark Swain. Inaccessible Island can be seen behind. (Andy Repetto)

The pensioners' bus at its Patches Plain terminus looking east towards the Potato Patches, Hillpiece and Burnt Hill. This service is provided free for over-65s from the village. (Chris Bates)

imports of South African groceries. Now, young islanders were able to follow Premier League football teams and families would look forward to *Strictly Come Dancing* on Saturday nights, or follow soap operas such as BBC TV's *EastEnders*. Many tuned into ITV's *Good Morning Britain* to hear medical advice from former Tristan doctor Hilary Jones, the programme's resident medical expert. As this century has progressed people on Tristan have also enjoyed access to social media, and many have their own Facebook and Twitter accounts.

Islanders now travel on the Settlement Plain using a range of motor vehicles. The limited road network still only consists of paved streets in the village, a rough track over the 1961 lava flow to Big Point, and a partially paved road westwards up the Valley to the Potato Patches and onwards to the Bluff. The factory and government own a range of vehicles including tractors and trailers, an articulated lorry and a few JCBs, as well as several Land Rovers, for the use of the Administrator, and for police, fire, and ambulance services. Most families now have their own pick-ups, known by the South African name bakkies, which are convenient for travelling about and carrying loads of such commodities as potatoes or meat. Motor bikes are popular, and cycles are often used by children around the Settlement. Imported diesel or petrol is expensive, but a full tank lasts a long time on the modest road system with the longest round-trip being less than ten kilometres.

View showing one of several monitoring installations forming the CTBTO Tristan station on Hottentot Fence. (Maxime Le Maillot)

A striking feature of Tristan in the modern age is the plethora of high-tech monitoring devices maintained by the Vienna-based Comprehensive Nuclear-Test-Ban Treaty Organisation (CTBTO). Tristan's strategic position in the middle of the South Atlantic Ocean made it an ideal addition to the global network of 300 CTBTO monitoring stations in 2006. This brought an immediate benefit to the community as a 24-hour electricity supply was essential, and for the first time this improvement to life was extended to the whole village. The facility is operated by the French-based company Enviroearth on behalf of the CTBTO and staffed for most of the year by a series of French nationals who live in the village, supported by local staff. This wealth of sophisticated equipment provides data to detect any nuclear tests which were banned in 1996. It includes two seismic stations, one on Hottentot Fence, and the other at the Molly Gulch on the Patches Plain. Records are sent to Vienna where recordings made on Tristan help the world's continuing efforts to monitor and attempt to eliminate any atomic weapons.

The Tristan community has catered for visitors informally for many years, welcoming early ocean liners like the *Asturias* (1927), *Empress of France* (1928) and the *Duchess of Atholl* (1929) which brought the first tourists to step ashore. Islanders became aware that visitors would eagerly buy souvenirs such as woollens and they began to make curios like rockhopper tassel mats and model long. Even Prince Philip, during his visit in 1957, was presented with some marbles made from the eyes of bluefish for Prince Charles! Once Tristan stamps were produced, the Post Office offered them for sale to those aboard the visiting ships. The growth of luxury cruising from the 1960s resulted in regular calls by famous cruise ships like the *QE2* whose first visit in 1979 was marked by a special stamp issue. This 70,000-tonne ship was the largest vessel afloat when it was launched in 1967 and ships of that size could not land passengers at the island. So, Tristan officials went aboard, setting up a Post Office for the occasion, while others visited the ship with their 'trade-bags' of woollens, models and even 'floating stones'. These novel pumice pebbles which bobbed about on water were eagerly snapped up by passengers, keen to own a piece of the island to take home. It has been the growth of smaller expedition-cruise vessels, with their own RIBs allowing quick transfer of passengers ashore to Tristan and onto the outer islands, that provides a more sustainable tourist market as passengers are more likely to be able to achieve a landing. Nightingale Island can provide the highlight of visits when guides from the Conservation Department show passengers wildlife such as seals, rockhoppers, and mollies as well as the rare finches. Occasional visits to Inaccessible Island's Salt Beach are made with the hope of catching a rare sighting of the Inaccessible Island rail. Islanders know that keeping these islands in a pristine state will be the key to attracting more visits from passengers with a passion for the environment, and this means continuing vigilance to prevent the inadvertent introduction of plants and invertebrates.

Incoming mail for islanders is given out by hand after each delivery arrives by ship from Cape Town about nine times a year. This 2009 stamp features a photograph with Post Office Head of Department Iris Green handing a letter to Daphne Repetto with Kirsty Green (now Repetto) alongside. The counter is situated in the Prince Philip Hall where people gather after the mail gong is rung to wait for their name to be called and collect their precious letters and parcels. (Tristan Post Office)

The Tristan Post Office continues to produce regular issues of attractive postage stamps but, as stamp collecting has steadily become less fashionable, sales are far less that those achieved in the peak years of the 1980s. A stark comparison shows that stamp sales then averaged £250,000 per year but by 2014-18 they had declined to £55,000 annually. So, sales have fallen by a huge 78 per cent, even without considering inflation of 300 per cent in the same period. Fortunately for the Tristan Post Office, royalties from the sale of souvenir coins, organised by specialist companies in the UK, have averaged £165,000 in the five years to 2018. Whereas production and handling costs of stamps need to be deducted from the stated income, coin royalties are received as net profits.

Iris Green became Head of the Post Office in 2007, and in 2009 a new Tourism Department was created with Dawn Repetto as its co-ordinator. The pair worked together to plan a new combined Post Office and Tourism Centre with construction starting in November 2009. The new facility, including a museum, gift shop and café opened in March 2010. The partnership between the Post Office and Tourism Departments created a team that could work flexibly together under one roof and was the catalyst for a series of initiatives.

Tristan's oldest-ever islander Ellen Rogers at the stable door of the Thatched House Museum in 2013. The house was built using traditional materials by pensioners in a project co-ordinated by Head of Tourism Dawn Repetto. The large, yellow-coloured stone blocks are of volcanic tuff quarried with axes from a source at the Goat Ridge. The thatch is New Zealand flax which replaced the original tussock grass roofing material after it was introduced in the 19th Century. (Lorraine Repetto)

The museum expanded to exhibit a range of artefacts and in 2012 a separate Thatched House Museum was constructed by older islanders using traditional local materials. Its gable ends were built of soft-stone and the roof was thatched with New Zealand flax, thereby keeping alive skills that could be lost now that houses use modern materials. In 2016 the island was successful in being awarded a grant from the British Library for a pilot project in its Endangered Archives Programme entitled '*All hands' things*: the endangered archives of Tristan da Cunha'. Since then, modern scanning and storage equipment have been acquired and a start has been made to digitise valuable original documents held on the island. The coffee shop was re-named the Café da Cunha in 2013 and the centre became a regular venue for small events as well as a hub for visitors during cruise-ship calls. Tourists have a range of organised activities to choose from: trips to the Thatched House Museum, golf on the Hottentot Fence 9-hole course, a Geotrail following historical sites around the village, and hikes up the 1961 volcano, to the Potato Patches (with a bus trip option), or mountain Base. Craft displays and tours of the fishing factory are also available. The island football team, the Tristan Tigers, holds occasional matches against teams from visiting ships, though they have yet to enjoy the pleasure of an away game!

The cruise ship MS *Bremen* off Tristan on Boxing Day 2019, photographed from the Residency with the Administrator's flag flying on its customary garden pole. (Sean Burns)

In a typical year there will be about six visits by cruise ships, all advertising the Tristan days as provisional since landing is not possible in high seas. Regular ships include the Oceanwide Expeditions' MV *Plancius,* which carries out a late-season re-positioning cruise from Ushuaia in Argentina via the Falkland Islands, South Georgia and onwards to St Helena, Ascension and the Cape Verde Islands. More typically cruise ships start from Ushuaia and arrive at the Tristan islands from South Georgia but continue to Cape Town. Such vessels include the Dutch training sailing ship *Bark Europa,* the Silversea cruise ships *Silver Explorer* and *Silver Cloud*, and the French Ponant cruise liner *Le Lyrial*.

Special cruises aboard the RMS *St Helena* in 2006 and the MV *Island Sky* in 2011 enabled passengers to stay ashore in guest houses or island homes, in a rare boost for local tourism relying on day visits from cruise passengers. Recent use of large ship tenders, combined with a newly deepened harbour, meant that more people could come ashore, whilst the maiden voyage of the MS *Seabourn Sojourn* in January 2018 allowed for a sizeable group of 400 passengers to visit for what proved to be an exceptionally busy day for the Tourism Department.

The only assured way to arrange a holiday on Tristan da Cunha is to use one of the scheduled ships that ply to and from Cape Town to the island. Straightaway there is a considerable hurdle to mount in that a would-be visitor must first apply to the Island Council for permission to stay, citing the reason for their visit. Islanders remain wary of filmmakers or writers who may pry into their private lives after press intrusion in the 1960s and more recent misrepresentations. So, applications from the media are carefully scrutinised. Somewhat dramatically, the best-selling author Simon Winchester has been banned from re-visiting the island after writing insensitive remarks in his 1985 book *Outposts* following a visit in April that year. Therefore, he was not allowed ashore in 2000 when lecturing on a visiting cruise ship, or in 2009 when arriving on a

Post Office and Tourism staff ready to receive guests from the MS *Bremen* on Boxing Day 2019.
From left to right - front: Rachel Green, Dawn Repetto, Iris Green;
back: Shirley Squibb, Kelly Green, Caryn Green, Jane Repetto and Lillie Swain. (Sean Burns)

motor yacht. During a recording he made while marooned offshore aboard the *Corinthian II* for the BBC's radio programme *From Our Own Correspondent*, Winchester sympathised with the islanders' stance. He concluded that he had been given a lesson in the ethics of tourism and that the 'people of Tristan, in obliging me to stay away and remain here (offshore), were quite probably ... absolutely right.'

Once permission is granted, would-be visitors must attempt to obtain a berth on a ship. The fishing company vessels carry twelve passengers. By law, a vessel conveying more than that number must carry a doctor, so even exceptionally large cargo ships only cater for a maximum of a dozen travellers. Nearly all these twelve berths are occupied by essential travellers, usually expat workers and their families, and returning islanders. Summer voyages are particularly busy, but a single traveller or a couple may obtain berths, especially if they can be flexible with dates. The annual *Agulhas II* voyage offers about forty berths to the island travelling out in September and returning in October. Unfortunately, owing to the voyage's popularity with expat workers and island families there may be only a handful of remaining places for tourists and there is a very long waiting list.

Visitors have the choice of hiring a guest house, being able to opt for self-catering or to have meals provided by the owners. A popular choice is to stay full board in islanders' homes, which is a way of becoming part of the community and sharing in the routines of island life. With careful planning and patience, a holiday on Tristan is possible. Those who enjoy getting on with people, have an interest in outdoor life, and are prepared to join in with the activities available will have the experience of a lifetime, as many fit expat workers with a love of hiking and wildlife have discovered.

Tristan English remains distinctive and something that newcomers will find fascinating. The strength of this local dialect varies between individuals and age groups, but it has continued to develop and in some respects it may have intensified. There are some striking examples. 'H's are often added to words starting with a vowel; the Administrator is therefore called the *H'admin*, asthma may be pronounced *h'asmere*, and people might be seen 'heating heggs and happles'. The word 'done' is over-used, so phrases like 'I done went' or 'she done sent' will often be heard. 'V's frequently become 'W's; so, reference to the 1961 *wolcano*, and *Wera* (for Vera) will be noticed. Even more intriguing is the fact that flour often becomes *flubba*, shower *shuba*, and hour *hubba.* Visitors need to know that being 'touched up' on the island means being under the influence of alcohol (being tipsy); otherwise, the turn of phrase might be the cause of unwarranted alarm.

2010 view of the Settlement from the Base edge during the visit of MV *Corinthian II* which can be seen anchored off the 1961 lava flow. (Dr Carel van der Merwe)

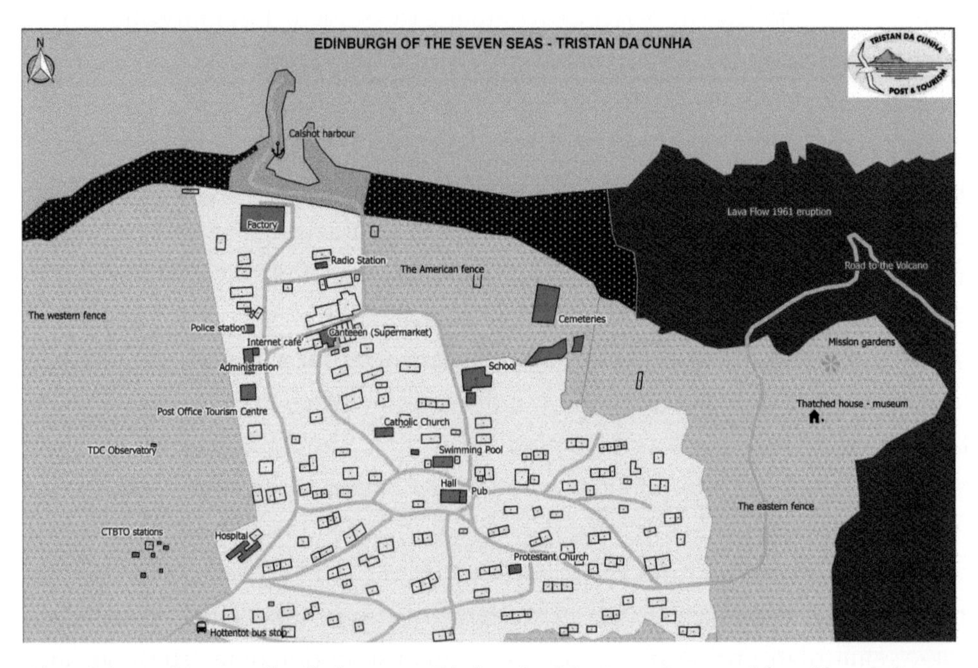

Plan of the Settlement, officially known as Edinburgh of the Seven Seas, used for self-guided tours for visitors to the island. (Tristan Tourism Department)

The Tristan dialect and language are now better understood thanks to the studies of the Swiss Professor of Linguistics Daniel Schreier, who explored Tristan English, discovering that the main influences on the language were the English spoken by early male settlers and by the group of women who arrived from St Helena in 1827. Schreier could find no examples of Scottish phrases or intonation as might be expected, since the founding father William Glass was brought up in Kelso. He concluded that as Glass's parents had moved from London, it was more likely they spoke a version of Cockney than the typical Lowland Scottish English prevalent in the Borders. Schreier attributes many grammatical changes (e.g. the use of 'done') to the St Helena women who introduced elements of their Creole language and were the main educators of the children at the time. He also notes that the well-educated Dutchman Peter Green, like the Italian Andrea Repetto who followed him, both spoke competent English and consequently there are no obvious Dutch or Italian words in regular use on the island.

Health and population matters

An old claim on Tristan was that islanders usually die of accident or old age. The island is indeed a dangerous place, and since resettlement in 1963 there have been six fatal accidents. Norman Swain was killed when struck by a falling rock while resting with his wife Maud above Plantation Gulch in 1964; Matthew Repetto was drowned at sea in 1968; Joe Glass, member of the Royal Society Expedition in 1962, disappeared when he was believed to be hunting for nightbird eggs in caves under the cliff towards Hillpiece in 1969; Bernard Repetto died while fishing for Tristan lobster in a factory boat in 1980; Barry Swain was swept out to sea and drowned when walking along the narrow beach with three other island men between Anchorstock and the Caves in 2012; and Alfred Rogers was killed in the island's first serious road traffic accident when he lost control of the tractor he was driving and crashed on the steep harbour road in 2017 – a saddening indication that, though isolated, islanders also face hazards similarly encountered elsewhere on a daily basis across the world.

Unfortunately, the relatively good health of the old community, characterised by fitness and a diet high in protein but low in sugar, is now a thing of the past. In the 21st century islanders have a typical western diet with imported foodstuffs readily available at affordable prices in the store; so, the consumption of fat, sugar and alcohol is higher than was traditionally the case. With UHT milk readily available in the shop, few cows are now milked for local consumption. Islanders are good cooks and buffet tables for birthdays and weddings are adorned with delicious home-made cakes and savouries. Poorer diet allied to a more sedentary lifestyle, with vehicles often used for journeys around the village, mean that many islanders are not as fit as their ancestors. Many are overweight, and a longer life expectancy is threatened by the inevitable health problems which consequently arise. Nevertheless, the island's oldest-ever resident Ellen Rogers was 102 when she died in 2021, providing an example of healthy living to the younger community.

In recent years, the UK Government has invested heavily in Tristan's health, most notably in providing £8.3m funding for the construction the impressive new Camogli Healthcare Centre, which replaced the old Camogli Hospital in 2017. The UK company Galliford Try used prefabricated sections manufactured in Sweden to complete the building in just eight months. This striking facility would be the pride of a British town twenty times bigger than the Tristan village. There is an emergency treatment room, laboratory, radiology room and well-equipped dental suite. There are two wards for in-patients, with two beds in each of them, and beds in two consulting rooms which could be used in an emergency. There is also a spacious family and waiting area with a library where staff can make hot drinks. There is a large main theatre, a preparation room, two changing rooms for staff, an instrument storage room, and a sterilisation unit. As well as providing the capital for the new facility, the UK Government funds expatriate staff, with a regular complement of two qualified doctors and two nurses. Dental and optical specialists visit regularly. This results in an enviably high ratio of healthcare professionals for a resident population of under 250 people, with care immediately available to all who live within the tiny village.

Patients still need to travel to South Africa and occasionally to the UK for specialist treatment. These 'medevacs' usually have an islander relative as an escort and can use Tristan House in Cape Town as a base for out-patient appointments. Health care is a major expense for the Tristan Government, amounting to an annual expenditure of over £250,000 for the five years up to 2018 without including the salaries of expat doctors and nurses.

Tristan's population grew strongly in the 1970s, beginning with a post-volcano birth boom and the return of many who had gone back to England. During the 1980s there were a record 300 resident islanders. Nevertheless, mainly young people have left Tristan, often travelling to the UK where islanders have right of residence and work. This migration was temporarily halted in 1981 when the UK Nationality Act denied Tristan islanders full rights of abode as British citizens. A relaxation of the law in 1998 allowed three island girls to live and work in England, and in 2002 the British Overseas Territories Act returned the right to hold full British passports to islanders once again. Forty islanders have departed and not returned since 1963, 27 of them (67.5 per cent) being women.

Inbreeding has been an unspoken concern on Tristan for some time, but this is a problematic subject for a close-knit community. The island's population grew out of the strength of an unusually diverse range of settlers from Great Britain, the Netherlands, Italy, the USA, St Helena, and South Africa, giving the island a much healthier genetic mix than most other settled communities. Any baby who has a non-islander as one of their parents is welcome in a society concerned about inherited health problems. Young people, who fail to find a partner who is not a close relative, may leave the island to seek a husband or wife aboard. Successive Island Councils have discussed easing the highly restrictive immigration policy which grants permanent resident status only to those born on the island and to their respective spouses and dependents. New arrivals could provide new skills and, if families settled, this would help alleviate the hazards of inbreeding. In the 21st century two new settlers have arrived, both qualifying for resident status by marriage.

Shirley Swain married Jody Squibb on Tristan in February 2010 after which the couple returned to England, but in 2012 they decided to settle on Tristan, with Jody gaining full residency status two years later. Jody became the first male settler to have remained on Tristan since the Italian sailors, Lavarello and Repetto in 1892, and the Squibb surname has therefore been added to the Tristan community. The couple now have two children. The second arrival was Kelly, the daughter of Administrator Sean Burns and his wife Marina. Kelly visited her parents in 2012, and during her time on the island, met her future husband Shane Green. The couple were married in September 2017 during Burns' second term as Administrator. Kelly became the first woman to settle on Tristan since the Irish sisters Agnes and Elizabeth Smith in 1908. The couple also have two children. It was significant that Kelly was elected to the Island Council in 2019, receiving more votes than any of the other women candidates. Sean became the longest-serving Tristan Administrator when he completed his second three-year term in January 2020.

Gladys Lavarello took over the role of island midwife from her mother Mary Swain after completing a midwifery training course on St Helena in 1972. She delivered her first Tristan baby in June 1973 and after her 'retirement' in 1996 continued her valued service, assisting with births until, at the age of 76, she helped deliver her grandson Ryan in 2007. Since 2012 mothers travelled to give birth in Cape Town, seeking specialist maternity care. However, with the opening of the new health centre, and specialist equipment and a suitably qualified doctor on hand, births resumed on the island when baby Alfie Rogers, was delivered on Tristan in August 2020.

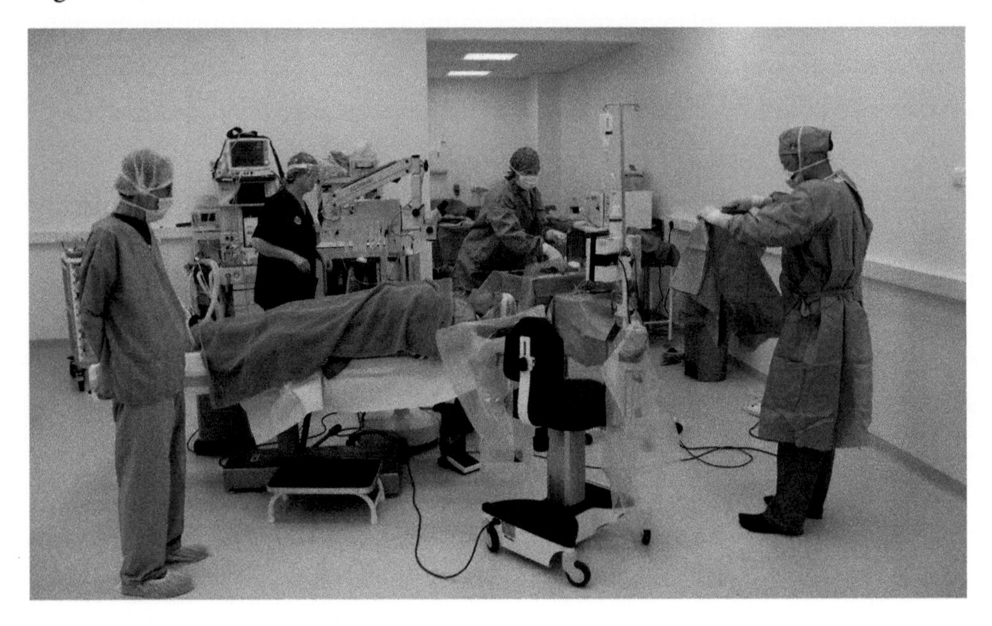

Operating theatre in action at the Camogli Healthcare Centre. (Dr Alex Wonner)

Tristan da Cunha islanders have long been known to suffer from asthma, and therefore are more susceptible to any viral infections that may cause bronchial problems, as was seen during their winters in England. Dr Noe Zamel, then a British Council Scholar, worked with the islanders at Calshot and discovered that over 50 per cent suffered from asthma, the highest known proportion anywhere in the world. Dr Zamel, now based in Canada, resumed his studies when he visited Tristan in 1993. Searching for a genetic cause for asthma, he traced five original settlers who carried the disease: sisters Maria Williams and Sarah Jacobs, and Sarah's daughter Mary Jacobs who arrived in 1827, and the Italian settlers Andrea Repetto, who died of asthma aged 44, and Gaetano Lavarello. So, a third of the original fifteen founding settlers suffered from asthma, accounting for the unusually high modern incidence of the condition. Dr Zamel took DNA samples for analysis which led to the crucial identification of the gene ESE3 which is a key trigger for asthmatic symptoms. Dr Zamel returned to

Tristan with a BBC *Horizon* team in 2008 to make a film which was broadcast the following year, and which featured Darren and Miranda Repetto and their son Randall, who was diagnosed with severe asthma as a baby. Dr Zamel also collected more DNA samples, hoping to identify further asthma-related genes which may in future result in a cure for the disease.

The 2020-21 Coronavirus pandemic was a challenge to an island which, with an aging population and a high incidence of asthma, was particularly vulnerable. The 2019-20 cruise ship season ended abruptly with the successful visit of MS *Bremen* on Boxing Day 2019, after which the island halted all visitors coming ashore as the pandemic struck. No cruise ships called in the following year while the world-wide crisis continued. Would-be visitors on yachts were prohibited from landing, and berths on scheduled ships were restricted to essential travellers. Strict quarantine arrangements were in force before departure from South Africa and on arrival at Tristan. A group of six islanders staying in Tristan House in February 2020 had to wait until September that year to return home as one of them tested positive for Covid-19, and, although asymptomatic, the whole group was refused permission to return earlier. Travel restrictions also meant that expatriates were marooned on the island beyond their contract term while others could not travel out to the islands to work. Delivery of vaccines posed a big logistical challenge which proved impossible by usual air and sea routes via South Africa as low storage temperatures could not be assured. Instead, the Foreign, Commonwealth and Development Office arranged for the supply of AstraZeneca vaccines to be flown to the Falkland Islands and then transported by the Royal Navy aboard HMS *Forth* to the island. The vaccines arrived on 21st April 2021 ready to be given to the population in the Camogli Healthcare Centre during the following weeks.

To be successful, Tristan needs a viable community with enough workers to keep the economy functioning effectively. The population has been in decline for several decades as the number of deaths exceed births and more people leave than arrive. From the peak of 300 in the 1980s the population has steadily fallen, which has meant that by the end of 2020 the figure stood at 244, with six of the islanders expecting to continue to live overseas. Attention has been already drawn to the shortage of labour to carry out construction work; it is a sobering reality that the community simply cannot function independently today without employing expensive expat labour paid for with grants from the UK Government.

Tristan enhances its international standing.

Tristan da Cunha's coat of arms (Tristan Government)

Under a Royal Warrant granted by HM the Queen in October 2002 a coat of arms was created for Tristan da Cunha and this was amalgamated into the territory's new flag. The crest shows a Tristan longboat above a naval crown, with a central shield decorated with four mollies and flanked by two Tristan lobsters. Symbolically these emblems together gave the island community its own official identity and paved the way for a new constitution seven years later. Formerly part of the UK's overseas territory of St Helena and Dependencies, the entity was renamed St Helena, Ascension and Tristan da Cunha with effect from 1st September 2009. Even though the territory's Governor is resident on St Helena, the new constitution gave equal weight to detailed separate provisions for Tristan da Cunha, which included a Bill of Rights for its citizens.

The new constitution clarified the continuing role of the Tristan Island Council to give advice to the Governor (through the resident Administrator). The Governor is not obliged to follow this advice, but if the holder of that office acts contrary to

Council's recommendations, any Councillor has the right to submit his or her views on the matter to the Secretary of State in London. In practice, the Governor has indeed considered the views of the Island Council and can now take part in meetings remotely via an internet link. It is, however, important to note that the Island Council remains an advisory body, as its counterpart also does on Ascension Island, though this is not the case on St Helena where the elected council also has executive powers. In the egalitarian Tristan community, there appears to be no appetite for islanders themselves to have more control.

The Residency, home of the Administrator with the Administrator's flag flying as usual from the flagpole during residence. It is the UK Government, through the FCO - rather than the Tristan Government - that pays for the upkeep of the Residency and the Administrator's salary and expenses. (Bernard Pronost)

Conrad Glass replaced Albert Glass as the island's policeman in 1989, becoming a Sergeant in 1993 and elevated to the rank of Inspector of the Tristan police force in 1998. Conrad is assisted by Special Constables who are called out of their offices for passport control when a cruise ship arrives. Those seeking an insight into Tristan life through the eyes of the law should consult his charming 2005 book *Rockhopper Copper* which confirms that some crime does occur on the island. As recently as 2020 a considerable theft was made from the Island Store, but the stolen money was returned anonymously. Conrad was Chief Islander between 2007-10. In 2008 the Tristan

Government appointed Chris Bates, who had helped Conrad to edit his book, as its first UK Representative, and he pioneered this new role which considerably enhanced the island community's links with the UK Government and other partners. Under Bates's stewardship, Chief Islanders Conrad Glass and Ian Lavarello attended Overseas Territories conferences and strengthened the relationship with Foreign Office and Overseas Territories ministers. This development enabled Tristan affairs to be separately considered as the island's more independent status took shape. Chris Carnegy took over the role of UK Representative in 2014 and since then he has robustly advanced the Tristan Government's external relations, working with Chief Islanders Ian Lavarello, who served three consecutive terms between 2010 and 2019, and James Glass, who was elected in 2019 for a record fourth term of three years.

Tristan Government UK Representative Chris Carnegy with Administrator Sean Burns at the FCO in July 2017. (Chris Carnegy)

With the arrival of the internet to the island the TDCA has been able to develop closer contact with Tristan. Islanders and expat workers, led by successive Administrators, have been able to submit up-to-date news and photographs of events for publication. The twice-yearly newsletter greatly extended its scope and since 2004 every island family has received its own complimentary copy. The TDCA also established a joint Tristan Government and Association website which was launched in 2005, and the site has continued to expand as information and news pages provide an increasingly comprehensive service to the community. Its topics include the marketing of stamps, souvenirs and handicrafts, advertising job vacancies, and promoting Tristan's crucial fishing industry.

The TDCA first organised a residential Annual Gathering weekend in 2005, and since 2007 these have all been held in Southampton, a location particularly convenient for those islanders living in UK, many of whom are based in Hampshire. In 2006 the Association organised a series of events to celebrate the quincentenary of the discovery of the island. These included a House of Lords reception, a special cruise to Tristan da Cunha aboard the RMS *St Helena*, which enabled members to stay ashore for six nights, and a dinner at the Royal Geographical Society attended by the Duke of Edinburgh. Also, in 2016 the Association established the Tristan Education Trust Fund, a UK-registered charity, and in 2020 it adopted a new constitution for itself. Fundamentally, the TDCA exists to support the Tristan community, and this is manifested by its on-going education grants and the donations it makes in times of need and to develop amenities.

HRH Prince Philip Duke of Edinburgh with members of the Tristan da Cunha Association at a dinner to celebrate the quincentenary of the island's discovery in 2006.
Left to right: Vice President Sir Martin Holdgate, Vice-Chair Lorna Lavarello-Smith, Prince Philip and Chairman Michael Swales. Lorna was the first Tristan student to benefit from a scholarship to Denstone College. She was ordained as an Anglican priest in the UK and in 2021 was continuing her ministry in South Africa. (Ian Mathieson)

Poisoned bait ready to be dropped by helicopter onto Gough Island to cull invasive mice during a trial project in 2015. (Ben Dilley)

A further TDCA initiative has been the introduction by Webmaster Peter Millington of Tristan traditional dances at the Annual Gatherings. On those occasions he has led a band to perform the music, and called the steps for the various items, including the famous Pillow Dance to form a popular feature at the event. Peter returned to Tristan in 2019 when he was invited to play his accordion to accompany traditional dances in the Prince Philip Hall.

It is in the field of conservation that the Tristan da Cunha islands are of global importance. Gough Island was declared a UNESCO World Heritage Site in 1995 and the site was extended to include Inaccessible Island in 2004. The islands are recognised by UNESCO as 'having one of the most pristine environments left on earth'.

Their world status is particularly enhanced by endemic seabirds including gonies, Tristan skuas and Inaccessible's spectacled petrel or ringeye, and land birds including the Tristan 'starchy', Gough finch and moorhen and the Inaccessible rail and finches. A Tristan Natural Resources Department headed by James Glass was established in 1995 and in 2008 it was divided, with James heading a Fisheries Department and Trevor Glass leading a separate Conservation Department. Links with the RSPB were strengthened, and many conservation workers have carried out field activities across the Tristan archipelago specially to study and monitor key sea and land birds.

An alarming discovery was made at Gough Island in 2000/01 when it was reported by scientists working there – and then by the world's press - that the feral house mice, probably introduced by sealing gangs in the 19th century, were causing the death of Tristan albatross chicks; later the chicks of several other species including mollies, sooties and Atlantic petrels were also found to be victims. The house mice of Gough Island are the largest known in the world and extremely abundant. They clearly pose a threat especially to a number of endangered seabird species and hence to its status as a World Heritage Site. The RSPB organised the raising of over £9 million to eliminate the invasive house mice from the island by dropping poisoned bait upon its surface by helicopter, with special arrangements to safeguard the unique native land birds. The Covid pandemic caused the project to be curtailed in 2020, but it has been resumed in 2021 and, if successful, it will be a huge boost to the island's wildlife, sustaining it as one of the most important seabird colonies in the world.

The Tristan Government, working with partner conservation organisations including the RSPB, the Marine Management Organisation, and the Centre for Environment, Fisheries and Aquaculture Science, has enthusiastically taken up the UK Government's Blue Belt initiative. The Pristine Seas *National Geographic* Expedition in 2017 launched the project, and further important research was carried out by expeditions aboard SVS *Grenville* in 2017, RRS *James Clark Ross* in 2018, and RRS *Discovery* in 2019. These built up a data base of the marine environment surrounding the Tristan da Cunha islands, with a particular focus on the seamounts.

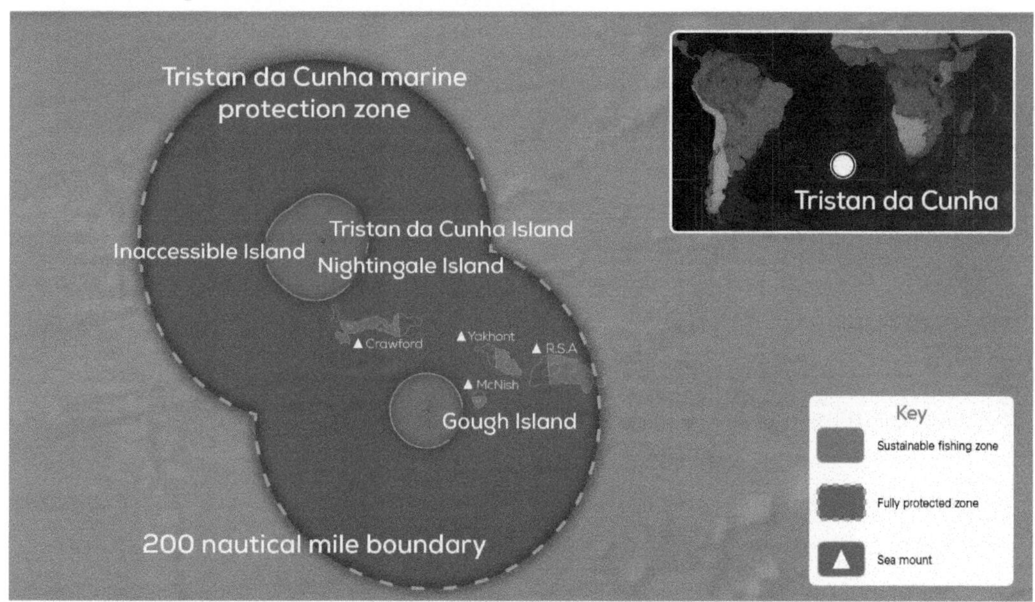

2020 Marine Protection Zone Map showing the relative position of the main four islands of the Tristan group, and archipelago's position in the South Atlantic within the inset. (Tristan Government)

A Tristan da Cunha Marine Protection Zone (MPZ) was formally announced in November 2020 and was welcomed across the globe, a rare positive media highlight in a year dominated by the dismal Covid-19 pandemic. At 687,000 square kms, Tristan's MPZ is the largest marine reserve in the Atlantic Ocean. Comprising around 91% of the islands' Exclusive Economic Zone, it provides a no-catch area for fishing. The rest of the EEZ is restricted to strictly controlled activities, mainly by the islands' licensee for sustainably caught Tristan lobster. There is scope for limited line-caught fish, but a total ban on bottom-trawling. Therefore, Tristan is at the forefront of the UK's Blue Belt plan to protect the world's oceans against unsustainable human use.

Traditional activities and values adapt to modern times

Sailing to Nightingale Island in traditional longboats in 1983. (Richard Grundy)

This final chapter looks at those special features of life on Tristan da Cunha which were sorely missed during their 1961-63 sojourn in England. There is no doubt that most of the islanders dreamed of being back in their beautiful island, enjoying the freedom and flexibility of working for themselves. Their thoughts would have turned to their beloved Nightingale Island as they endured cold winters in England. However, the tradition of hunting and gathering trips to Nightingale Island has markedly changed over the intervening years. The island's traditional wood-and-canvas longboats were maintained in excellent order by their owners for several decades, but the switch was made to fibreglass construction in the 1990s and their use was soon superseded by powered boats for the 49km trip to the island. The dull fibreglass longboats form a sad sight alongside the Residency wall near the school as their shells deteriorate, a seemingly poor substitute for their brightly painted predecessors. Inshore vessels are now of high quality, RIBs operated by the Conservation, Fisheries and Police Departments having the capability to make a return journey to Nightingale Island with ease but taking fewer people and less cargo than the fleet of longboats.

Ernie Repetto in front of his Nightingale hut in 1983. These huts were destroyed by the 2001 storm and few have been replaced. (Richard Grundy)

On Nightingale there exists a wilderness devoid of rodents and teeming with millions of breeding birds, where islanders continue to hunt and gather seabirds and their eggs. The taking of a limited number of rockhopper eggs in September and large numbers of petrel chicks for food and cooking fat in May is still permitted under revised conservation laws introduced in 2006, but trips have been much reduced as little guano is collected for fertiliser and the 2001 storm destroyed most of the huts which used to provide shelter. Some new huts have been built and the Conservation Department built a well-equipped base on the island, but Nightingale can no longer cater for the large numbers who joined longboat trips in the 1980s – sometimes they exceeded 100 people – nor provide overnight accommodation for most of the population as it did on the night of 10th October 1961.

Rockhopper and chick in the Nightingale tussock (Antje Steinfurth)

Well into the 21st century the community maintains its core way of living off the land as equals, despite changing trends in other aspects of the island's life. An Agriculture Department, supported by visiting expat vets and agriculture advisers, organises the husbandry of stock owned by islanders and grazed on common land. The rock-strewn pastures are poor, based on highly acidic soils leached of nutrients by heavy rain and dominated by low-grade species such as Kikuyu grass. Importing and applying fertilisers and lime can be costly, which means that improving the grazing is no easy feat. In 2019 the Island Council reduced adult cattle to one per family, in order to increase grazing availability and herd quality, and to prevent winter losses through malnutrition. Two adult sheep are allowed per person, whilst families keep large numbers of hens and ducks which recycle kitchen waste for egg production. The shortage of adequate pasture was a key reason why the island was considered overpopulated. Now that measures have been taken to limit stock, and the island's population is also falling, perhaps the Island Council will consider allowing some new families to settle.

Ducks graze behind St Joseph's Catholic Church. (Colette Halter-Pernet)

Since returning in 1963, the community's social calendar, mainly centred around farming activities, has strengthened. Their skills now also produce a wide range of handicrafts which reflect their culture. Islanders are seen to relish the traditional celebrations that make life on Tristan incredibly special compared to their more mundane life in England.

Sheep Shearing Day, usually held on a Saturday before Christmas, involves every family. The road from the village to the pens on the Patches Plain is jammed with a cavalcade of bakkies and other motor vehicles taking out people and refreshments for the day ahead. The scene in the pens has been unchanged for over a century: the sheep

– marked distinctively according to ownership – are gathered and shorn of wool by hand using steel clippers. This high-summer day is eagerly awaited by family groups of all ages who join in with the frenzied activity before gathering on the cropped grass to enjoy a picnic, though its real focus is on bringing home prized fleeces to be made into famous island woollens.

A busy scene on Sheep Shearing Day (Janice Hentley)

Those with the necessary strength work together in gangs to climb the mountain to tend sheep which breed and run wild on the rich pastures of the mountain Base. Just before Christmas, dogs round up the sheep which may have strayed right up to the higher cinder slopes below the peak. They are gathered in pens on the lower Base where they are shorn. Selected animals are killed, butchered and the meat carried back to the village. Mutton, stuffed with potatoes and a blend of herbs and spices is the traditional Tristan Christmas meal: mountain mutton, fed from plentiful grass on the Base has a stronger flavour and is a treat indeed.

Tristan woollens are still produced in the time-honoured way. After shearing, the wool is first washed and then prepared by 'carding parties' where it is meticulously combed out to align the fibres ready for spinning. Large metal spinning wheels, turned by hand, produce continuous threads which are washed and stored as skeins ready for knitting. Expats have attempted to introduce powered spinning and electric knitting machines, but these have been largely shunned by the conservative islanders. So, the village womenfolk can be seen daily (but never on a Sunday, at least in public) clicking away to produce another scarf, beanie hat, pair of socks or gansey, as jumpers are usually called. The enterprising Tourism Department now market Tristan woollens under the trade name '37 Degrees South' using a team of over fifty island women preparing bespoke woollens. A speciality is the 'Love Socks' range inspired by the tradition of indicating fondness for a man by the number of stripes knitted into the socks. Special orders have included an island wool patchwork screen for a multi-media exhibition in France, and bulk orders for an outlet in New Zealand where Tristan woollens are marketed as 'Queen Mary's Peak Knitwear'.

Hunting for rats with Head of Police Conrad Glass working a pole to open up another underground nest. Close at hand a border collie dog is braced to add to the collection of rodents visible on the ground to the left. (Janice Hentley)

In 2019 the Tourism Department launched a range of handmade jewellery and other products crafted from sterling silver, suede and locally sourced products like pebbles and flax. They chose the brand name 'Tesori di Tommaso', inspired by the Italian Tommaso Corri who lived on the island between 1810-16 and is thought to have left behind a buried treasure. Model longboats are also still built to exacting standards and are sought after by visitors.

Ratting Day is another distinctive holiday combining pest control with sporting competition. Teams are named after areas of the Potato Patches: Old Pieces, Second Watron, Below-the Hill, Coolers, Jack's Piece, Redbody Hill, Twitty Patch or The Farm. Groups set out at dawn with steel picks, shovels, and border collie sheep dogs. When the dogs smell a rat's nest in the earth below, there follows a frenzied dig. Picks upend grey lava boulders, shovels dig the earth, and dogs eagerly pounce to kill all the rats or mice that emerge, often at break-neck speed from their subterranean lairs. Gangs then retire to their Patches camping huts where hot food is prepared by their supporters, and drinks are passed around. The day is not for the squeamish. When the rodents are caught their tails are cut off (as close to their bodies as possible) to provide the evidence for the finale of the day: the counting of the tails. The resident doctor is always the official arbiter, and it was a tradition to use a table in the old Camogli Hospital for the

James Glass astride of a row of potato plants in one of his Patches. (Richard Moss)

count, but today health and safety concerns have even reached the South Atlantic, so it takes place elsewhere. Tails are counted, the longest tail (hence the careful cutting to the rump of the rat) is measured and the results logged. The teams gather in a tense atmosphere awaiting the results. Celebrations follow when the prizes are given out, and teams (unless they receive the Booby Prize) gain bragging rights to last another year. A splendid day of activity, and usually another occasion later for a dance in the Prince Philip Hall. The use of poison-baited boxes around the village and the Potato Patches, maintained by the Agriculture Department, has significantly reduced the mice and rats in these areas. A bonus has been that Antarctic terns, known locally as kingbirds, are now found nesting on walls at the Patches.

Successful potato growing still lies at the heart of a Tristan family's way of life. Although potato cropping for many years on the same ground is frowned upon by experts, islanders get good yields with the ingenious use of local fertiliser, a specialised menu which may include guano, lobster shells and offal from the factory, poultry manure and even sheep fleeces not needed for woollens. Other vegetables and salad crops are grown in sheltered gardens around village houses, and by the Agriculture Department in poly-tunnels at the Mission Garden, but the island relies on imports of staples like onions that are sold in the Island Store. Many islanders fish for their own food as well as for the fishing company, so home freezers may well have five-finger, bluefish or mackerel caught from boats or rods and lines from the shore.

Employment hours have evolved to suit islanders who want to tend crops and stock at the close of their paid work. So, after an early start, work continues without a break until the set finish-time in the early afternoon, which is never later than 3pm and 1.30pm on Fridays during summer. Fishing days are heralded by the sound of the gong before dawn and part-time fishermen transfer from government to factory-paid employment, which continues until the catch is safely brought ashore. Many other islanders, including seniors who remain keen to supplement their basic pensions, proceed to the factory for evening processing work on fishing days, but late-night processing is now curtailed as modern lobster storage allows this work to continue the following day. 'All Hands Days' are also declared when there is an absolute priority to get an activity completed swiftly, such as offloading a ship of crucial fresh food and groceries or making urgent repairs after a storm. On these days, the community pulls together towards a common goal with all the people working flexibly as a single team.

Queen's Day has thrived and grown in modern times as a mark of Tristan's continued loyalty to the British crown. The day is held in late summer after the crops have been gathered, though it is often delayed until good weather can be assured so that it can be celebrated in style. A show of produce, cookery, crafts, and children's art is assembled in the school with keen competition especially in the cake and potato classes. There is usually a beach-fishing competition, using slick imported rods and lines now that handlines have been phased out. In the afternoon competitions are run. For children there is a selection of egg-and-spoon, sack, and three-legged races; adults may take part in welly-wanging, javelin-throwing, keenly contested tug-of-war contests and football-rounders for all ages. Afterwards the Administrator holds a reception on the Residency lawns when prizes are given out and the Queen's health is toasted. Finally, a dance held in the Prince Philip Hall to round off a right royal day.

Anniversary Day on 14th August commemorates the occasion in 1816 when the garrison aboard HMS *Falmouth* took possession of Tristan da Cunha on behalf of King George III, and from that day forward the island fell under the protection of Britain. Apart from a special 200th anniversary party in 2016, this occasion is now usually marked by a quiet day off work, since its date is fixed, and outdoor events are difficult to organise in winter.

Tristan's coat of arms bears the motto: 'Our faith is our strength'. Christian faith and the role of the church have been integral to the island's history. There were frequent examples of this belief as islanders grappled with their refugee status but were firm in their conviction that God would protect them. On returning to the island, Lars Repetto led worship at St Mary's Church every Sunday morning for the fellow members of the Resettlement Survey Party and continued in a leading role there as Lay Minister, with Eddie Rogers and Carlene Glass-Green. In September 2019 St Mary's welcomed Rev. Margaret Van den Berg from South Africa, the latest in a long line of priests who have served the island. St Mary's Church has been modernised over the years, but it retains many historical artefacts, including the prized portrait of Queen Victoria, presented by the monarch herself to the island community.

The Roman Catholic community is also well-established on the island. A small church was built in 1983 and a handsome replacement dedicated to St Joseph was built in 1995/96. A whale-shaped wind vane adorns the western bell tower, and a stunning stained-glass porch window depicts St Joseph, in front of a snow-capped Tristan island, with a molly in flight and a longboat in full sail on an azure sea. Lay Ministers Dereck Rogers, Anne Green and James Glass, all grandchildren of the congregation's founder Agnes Smith have organised services, supported by an Abbot based in the UK who most years is able to make a formal visit to Tristan.

Stained glass window in the porch of St Joseph's Catholic Church. (Jenny Bowles)

Since their leaders are both men and women, the churches provide a useful prompt to reflect on equality on the island. Tristan is a society with equality of opportunity, where women play leading roles at parity with men. St Joseph's Lay Minister Anne Green was elected as the first female Chief Islander in 1988, served a second three-year term between 2003 and 2006, and also served as Head of the Education Department. St Mary's Lay Minister Carlene Glass-Green was the first woman to express the wish to work alongside men maintaining vehicles and is now Head of the PWD's Mechanical Department. Women have also led the Island Store, Finance, Tourism and the Post Office Departments. In addition, the constitution (unusually) requires that the

Island Council cannot be all-male. Reference is made in this book to 'fishermen', as indeed they are all currently men. Nevertheless, a young woman trained as an apprentice 'fisher' a few years ago, but she is one of those young people who have chosen to live overseas. It is pleasing to report that in 2020, Tristan welcomed its first lady Administrator, Fiona Kilpatrick, who jointly took up the post with her husband Steve Townsend, alongside its first woman priest. Equal opportunities for women are therefore increasingly recognisable in modern Tristan society.

Island councillors gathered in the Residency on the hand-over of administration from Sean Burns to Fiona Kilpatrick and Steve Townsend in January 2020.
Left to right: Paul Repetto, Ian Lavarello, Steve Swain, Joint Administrator Fiona Kilpatrick, Chief Islander James Glass, Council Clerk Geraldine Repetto, Rodney Green, former Administrator Sean Burns, Warren Glass, Dawn Repetto, Kelly Green and Carlene Glass-Green. (Steve Townsend)

Perhaps, when enduring cold winters in England, it was the thought of the mid-summer Christmas holiday of the island that was a key factor that encouraged islanders to return. Today, all government and fishing work shuts down for four weeks, except for a few core services or for a ship arrival. Break-up Day heralds the long festival as work teams gather for parties, often around a braai or buffet table. Before this, St Mary's School pupils perform a Christmas concert, to showcase the children's love of dressing up, acting, dancing and singing in front of their families. The two churches hold carol services and people arrive at St Mary's hours before Midnight Mass to ensure a good seat; Catholics can take their time as there is ample room for the smaller congregation at St Joseph's, which makes up about 20 per cent of the population.

Families have time to relax, clean their homes ready for the 'big eats', and put fresh flowers on family graves. New potatoes are dug, and stock is tended. Many stay out in evermore elaborate camping huts at the Potato Patches, now well-equipped and a home-from-home only 3km away, and yet a change of scene for the holidays. It is often hot at this time of year, and Runaway Beach below the Patches is a popular place for people to gather. Here they may do some beach fishing, swim in rock pools, and relax on the grey volcanic sand, often hot to stand on with bare feet or to touch in the strong summer sunshine.

Okalolies arriving up the Residency path on Old Year's Night 2006. (Janice Hentley)

Back in the village the usual bustle of commercial activity stops. Families must stock up in advance as the canteen is closed and the Albatross Bar, the island's only pub, is shut for the entire holiday period. Old Year's Night sets the scene for group of island men (it is assumed!) furtively dressing up as Okalolies in ever more elaborate guises and wearing ghoulish masks and costumes. Hoping to be disguised by their kit, these characters tour the village touting for drinks and generally larking around. In the evening they make a mass entrance at the Administrator's reception and later at a similar party hosted by the Chief Islander. At midnight people gather around the island gong in the middle of the village, today an old steel gas bottle hung from a frame to ring in the new year as they also do by chiming St Mary's Church bell.

The modern Tristan village may not appear as attractive as the view of thatched dwellings captured by the British artist Edward Seago in 1957, but Seago cleverly chose a site that completely hid the ugly Big Beach fishing factory complex. Today's village is dominated by unattractive factory and government buildings and the houses further up the slope have lost their rustic thatch-roofed cottage-like appearance. Yet the modern homes are far more comfortable to live in. They have a range of amenities comparable with their UK counterparts, though they have no need of central heating as frost is unknown in the village. All homes retain their stable front doors facing the sea, many with front gardens adorned with colourful flowers. All the houses have variously coloured zinc-coated steel roofs which look particularly attractive from the sea. It must also be borne in mind that the Tristan settlement is a farmyard, every house is a farmhouse, so border collie dogs scuttle around, cows graze in the village, and everywhere there is the to-and-fro of people with tools, crops and stock animals.

A male fur seal on a Tristan beach (Marthán Bester)

Way back in Tristan's past the Settlement Plain was cleared of its native vegetation and wildlife to provide land for farming. But wildlife is once again finding a niche as human habits change. Nowadays, as open fires have been replaced by gas for cooking, firewood is no longer cut from the cliffs, which are now green with island tree foliage once more. This cover provides nesting sites for Tristan thrushes and Gough moorhens (released onto Tristan in 1956) which are now seen in the village itself where imported bushes and trees have flourished. Antarctic terns breed successfully on hardies, and nests are now also regularly seen on Patches walls. Tristan skuas breed around the 1961 lava flow and the skies above the village are often filled with increasing numbers of elegant mollies soaring on rising air currents as they cast their wide circles to their mountain nests above. Fur seals are more readily seen than before on beaches near the village, while the breeding colony at Seal Bay is thriving with 75 pups born in a season. From December to March moulting rockhoppers come ashore into the village in increasing numbers and can be found sheltering in gardens and even making their way into buildings. These green shoots of a wildlife recovery would accelerate if rats and mice were eliminated from Tristan. Perhaps, with imagination, careful planning, and enough capital, a future Edinburgh of the Seven Seas could be the jewel in the crown of the UK Overseas Territories presenting as its most appealing feature attractive and functional buildings amidst a rich natural environment.

Approach from the sea on a clear day and the view of the Settlement will take your breath away. Above, a sometimes-snow-covered volcanic peak crater rim, over 2000 metres high, slopes down to the edge of the Base 800 metres above the sea. Immediately behind the village there is an immense and threatening grey slab of bare rock. To the right the cliff is split by the huge V-shaped gorge of Hottentot Gulch where rockfalls are a daily occurrence.

To the left there stands a squat black mound, topped by jagged stones – this is the 1961 eruptive cone – and below it the mighty mass of black lava hides Big Beach and the site of the old lobster factory. Many people may feel threatened by this setting, but the narrow shelf is home to the fellowship of people who defied the odds and returned to re-claim their home in 1963. They are still there, and it is the hope of the authors that this book begins to explain why they made the right call. Nothing could stop them.

View from American Fence tri-purpose playing field, cattle pasture and helipad showing the red-roofed St Mary's School, the white-fronted Prince Philip Hall and, in the background, Hottentot Gulch bounded by the sheer cliffs that rise so dramatically behind the village. (Peter Millington)

Reflections

Tristan's core values retain their influence in modern times. The equality established in 1817 by William Glass remains intact with all land still under communal ownership. This means that every islander has the right to build a home, grow crops and keep animals according to the law of 'fair shares for all'. Full employment is maintained through factory and government jobs, with equal opportunities offered through free education and access to training.

Co-operation continues as islanders give their labour freely to family and friends in a variety of ways. Families remain at the centre of a social life with the wider community often invited to attend christenings, special birthdays, receptions and particularly weddings which are by tradition public occasions of celebration. There is no doubt that it was these core values and practices which were the driving force behind the decision of the islanders to return to Tristan in 1963 to re-form their unique society.

Traditions thrive on modern Tristan as Peter Millington plays the accordion for a Donkey Dance while Herbert Glass teaches Kelly Green the steps – a photograph taken during a party in the Prince Philip Hall to celebrate Kelly and Kirsty Repetto's 30th birthdays in November 2019. (Peter Millington)

Most islanders rejected the affluence and opportunities available to them in Britain. Instead, they chose to defy the odds, face the risks, and re-establish their distinctive way of life as far away from any other people as it is possible to live. They had to tackle huge problems and the possibility of failure unless a harbour and factory could be built. Present-day islanders now have the best of all worlds. They earn wages high enough to allow them to buy luxuries that provide a lifestyle way beyond the imagination of their predecessors. They still produce and gather their own food, and they continue to enjoy generous UK Government support that they shouldn't take for granted. They live as a peaceful group of people in a spectacular environment, safe from widespread crime and external threats such as terrorism.

For the future, the community will face continuing challenges, and these include the answers to the following three key questions:

- When and where will the next volcanic eruption occur?
- Will there be enough working-age people to sustain a viable economy?
- Will the UK Government be willing to continue providing the financial support to maintain the harbour and other essential services?

The TDCA aims to work with the Tristan community to support its future success. It hopes those who have read through this book and found out for the first time about this extraordinary group of islanders will feel inspired to join the Association and receive its regular magazine-style newsletters. If that is a step too far, these same readers are encouraged to keep an eye on the website www.tristandc.com and to see how the world's most isolated society fares in the future, whatever the tests may be that they will have to face.

The flag of Tristan da Cunha (Tristan Government)

Glossary of selected local terms

bakkie	South African term used on Tristan for an open-backed pick-up truck
bog fern	*Blechnum palmiforme*; dwarf tree-fern extensive on the lower mountain Base
braai	South African term used on Tristan for a barbecue
canteen	Island Store, named after original naval canteen
dolos or dolosse	tetrapod concrete blocks used as harbour erosion protection
fence	a field enclosed by walls or fences, e.g. American Fence
flax	New Zealand flax; *Phornium tenax*; provides village hedges and thatching
fur seal	subantarctic fur seal; *Arctocephalus tropicalis*
gony ***or*** **gonies**	Tristan albatross; *Diomedea dabbenena*
gulch	Normally dry valley, forming radial steep-sided ravines on mountain
gutter	narrow, steep-sided stream valley
hardy ***or*** **hardies**	offshore sea stack, often in groups, forming a headland of hardies
island cock	Gough moorhen; *Gallinula nesiotis*; introduced to Tristan in 1956
island tree	*Phylica arborea*; forms dense thickets, was extensively cut for firewood on Tristan
kelp	giant kelp; *Macrocystis*; huge fronds fixed to the sea-bed offshore to form a kelp forest
lobster	crayfish; *Jasus tristani*; called crawfish on Tristan; marketed as Tristan rock lobster
molly ***or*** **mollies**	Atlantic yellow-nosed albatross; *Thalassarche chlororhynchos*
petrel	great shearwater; *Puffinus gravis*; hunted for food and cooking fat
pond	all lakes are called ponds. There are nine crater lakes on the Tristan mountain
road	all paths, even narrow ones, are called roads
rockhopper	northern rockhopper penguin; *Eudyptes moseleyi*; alternative local name: pinnamin
sea elephant	southern elephant seal; *Mirounga leonine*
starchy ***or*** **starchies**	Tristan thrush; *Nesocichla eremita*;
station	applied to expat houses, offices and workers, eg station fellas; after naval station

tussock	tussock grass; *Spartina arundinacea*; formerly thatching material
watron	stream, brook; probably a corruption of watering

Commonly used abbreviations

CTBTO	Comprehensive Nuclear-Test-Ban Treaty Organisation
FCO	UK Foreign and Commonwealth Office; became Foreign, Commonwealth and Development Office in 2020
GISS	Gough Island Scientific Survey
MSC	Marine Stewardship Council
PWD	Public Works Department
RIB	rigid inflatable boat
SAIDC	South Atlantic Islands Development Corporation
SPG	Society for the Propagation of the Gospel; became USPG in 1965
TDCA	Tristan da Cunha Association
WVS	Women's Voluntary Service; became WR (Royal) VS - in 1966

Selected Sources and Further Reading

Unpublished resources

Andersen, Ole Magnus Mølbak: Account of the passage of MV *Bornholm* to Tristan da Cunha 1963.

British Library, ref. Add MS 20902: 'Relação du Naos e Armadas da India …,' arranged under the heads of the commanders of the expeditions from Vasco da Gama in 1497 to Luis de Mendonça Furtado in 1653.

--------, ref. Add MS 43846: Tristan da Cunha, documents relating to the early history of the British settlement on the island, 1817-24.

--------, ref. General Reference 10493.h.27: Collection of newspaper cuttings and pamphlets on Tristan da Cunha 1886-1946 compiled by Douglas M Gane.

Broadbent, Heather: Account of experiences with Tristan islanders in Hampshire from 1962.

Burn, Ron: Philatelic and historical archive.

Craig, Lucy & Jenny: Newspaper cuttings from Surrey 1961/62.

Repetto, Lars: Diary for 1962/63 exploratory party to Tristan da Cunha.

Shaw, Bob: Diary and account of the Royal Society Expedition 1962.

Taylor, Robin: Archive prepared for Peter Wheeler and letters.

Wheeler, Peter: Scrapbook with photographs, notes, letters and interviews.

Book-length publications

Baker, Peter, *et al.*, *The Volcanological Report of the Royal Society Expedition to Tristan da Cunha, 1962* (Royal Society, 1964).

Barrow, K M, *Three Years in Tristan da Cunha* (Skeffington, 1910).

Brander, J, *Tristan da Cunha 1506-1902* (George Allen & Unwin, 1940).

Crabb, George, *The History and Postal History of Tristan da Cunha* (self-published, 1980).

Crawford, Allan B, *Penguins, Potatoes & Postage Stamps* (Anthony Nelson, 1999).

--------, *Tristan da Cunha and the Roaring Forties* (Charles Skilton, 1982).

--------, *Tristan da Cunha: wartime invasion* (George Mann, 2004).

Earle, Augustus, *A Narrative of a Nine Months' Residence in New Zealand in 1827, together with a Journal of a Residence in Tristan d'Acunha* (Longman, etc., 1832).

Evans, Dorothy, *Schooling in the South Atlantic Islands 1661-1992* (Anthony Nelson, 1994).

Falk-Rønne, Arne, *Back to Tristan* (George Allen & Unwin, 1967).

Flint, Jim, *Mid Atlantic Village* (Best Dog, 2011).

Gane, Douglas M, *Handbook of Tristan da Cunha*, new ed., (Preedy's, n.d. [1927]).

--------, *New South Wales and Victoria in 1885* (Sampson Low, 1886).

--------, *Tristan da Cunha: an empire outpost and its keepers* (George Allan & Unwin, 1932).

Glass, Conrad, *Rockhopper Copper* (Orphans Press, 2005).

Grundy, Richard, *et al.*, *Commemorative Publication to Celebrate the 500th Anniversary of Tristan da Cunha* (Tristan da Cunha Association, 2006).

Hicks, Anna, *An Interdisciplinary Approach to Volcanic Risk Reduction under Conditions of Uncertainty: a case study of Tristan da Cunha* (Ph.D. thesis, University of East Anglia, 2012).

Holdgate, Martin, *Mountains in the Sea* (Macmillan, 1958).

--------, *Penguins and Mandarins* (Memoir Club, 2003).

Hosegood, Nancy, *The Glass Island: the story of Tristan da Cunha* (Hodder & Stoughton, 1964).

Jackson, E L, *St Helena: the historic island from its discovery to the present date* (Ward Lock, 1903).

Kornet-van Duyvenboden, Sandra, *A Dutchman on Tristan da Cunha: the quest for Peter Green* (George Mann, 2007).

Lavarello-Schreier, Karen & Schreier, Daniel, *Tristan da Cunha and the Tristanians* (Battlebridge, 2011).

--------, *Tristan da Cunha: History, People, Language* (Battlebridge, 2003).

Lockhart, J G, *Blenden Hall: the true story of a shipwreck, a casting away and life on a desert island* (Philip Allan, 1930).

Mackay, Margaret, *Angry Island: the story of Tristan da Cunha 1506-1963* (Arthur Barker, 1963).

Milner, Rev. John & Brierly, Oswald, *The Cruise of H.M.S. 'Galatea', Captain H.R.H. the Duke of Edinburgh, K.G., in 1867-68* (W. H. Allen, 1869).

Munch, Peter, *Crisis in Utopia: the story of Tristan da Cunha* (Longman, 1971).

Newman, George, *Other Lyrics: an aftermath* (Kent County Examiner, 1900).

Perry, Roger, *Island Days* (Stacey International, 2004).

Report of the Scientific Results of the Exploring Voyage of H.M.S. 'Challenger' during the Years 1873-76, 50 vols (HMSO, 1885-95).

Rogers, Rose Annie, *The Lonely Island* (George Allen & Unwin, 1926).

Ryan, Peter, *et al.*, *Field Guide to the Animals and Plants of Tristan da Cunha and Gough Island* (Pisces, 2007).

Schreier, Daniel, *Isolation and Social Change: contemporary and sociohistorical evidence from Tristan da Cunha English* (Palgrave Macmillan, 2003).

Svensson, Roland, *Tristan da Cunha, South Atlantic* (Proprius, 1965).

Taylor, Rev. William F, *Some Account of the Settlement of Tristan d'Acuna in the South Atlantic Ocean* (Society for the Propagation of the Gospel, 1856).

Taylor, Robin, *Tristan da Cunha Monographs,* 20 vols (self-published, 2001-09).

Tristan da Cunha Fund Reports, 1921-64.

Wace, Nigel & Holdgate, Martin, *Man and Nature in the Tristan da Cunha Islands* (International Union for Conservation of Nature & Natural Resources, 1976).

Watkins, Brian, *Feathers on the Brain!: a memoir* (Hardwick Hill, 2011).

Wild, Frank, *Shackleton's Last Voyage: the story of the 'Quest'* (Cassell, 1923).

Winchester, Simon, *Outposts* (Hodder & Stoughton, 1985).

Magazine articles, etc.

In addition to numerous references in the *Manchester Guardian, The Times* and the Australian press (accessed via https://trove.nla.gov.au/) the following are worthy of note:

Blackwood's Edinburgh Magazine, Dec. 1818 (Alexander Walton, 'Tristan d'Acunha, etc., Jonathan Lambert, Late Sovereign Thereof').

British Museum Quarterly, Sep. 1934 (R Flower, '19. The Tristan da Cunha Bible'); May 1935 (R Flower, '80. Tristan da Cunha Records').

Bystander, The, 18 Sep. 1907 ('A Community of Crusoes').

Corona, The (Journal of HM's Colonial Service), Aug. & Sep. 1950 (D I Luard, 'Tristan da Cunha [1 & 2]').

Graphic, The, 9 Jan. 1897 (sketch of Peter Green); 12 Jan. 1907 (A G Wise, 'The Sad Plight of Tristan d'Acunha'); 12 Feb. 1921 ('The Loneliest Spot in the World').

Illustrated London News, 9 Feb. 1867 (sketch of H.M.S. *Galatea*); 12 Oct. 1867 (Oswald Brierly's sketch of Peter Green's house); 14 Oct. 1905 ('The "*Pandora*'s Box" Case'); 30 Aug. 1924 ('A New Flightless Bird: the rail of Inaccessible Island'); 21 Mar. 1925, ('Asking for an Annual Mail'); 5 Mar. 1932 ('Tristan da Cunha: perfect teeth despite broken rules'); 31 Mar. 1934 ('Ruled by the Chief Woman and her Son, Chief Man'); 3 Apr. 1937 'Tristan da Cunha to Yield Data about Droughts in South Africa?').

Journal of Social History, vol. 49, no. 1 (2015) (Lance van Sittert, 'Fighting Spells: the politics of hysteria and the hysteria of politics on Tristan da Cunha, 1937-38).

National Geographic Magazine, Nov. 1938 (Robert Foran, 'Tristan da Cunha: isles of contentment'); Jan. 1950 (Lewis, 'New Life for the "Loneliest Isle"'); Nov. 1957 (Beverley M Bowie, 'Off the Beaten Track of Empire'); May 1962, (Peter Wheeler, 'Death of an Island'); Jan. 1964 (James P Blair, 'Home to Lonely Tristan da Cunha').

Nature Communications, 11 Sep. 2020 (Wolfram H Geissler *et al*, 'Seafloor Evidence for Pre-Shield Volcanism above the Tristan da Cunha Mantle Plume').

Picture Post, 5 Nov. 1949 ('Tristan Begins a New Life').

South African Shipping News and Fishing Industry Review, April 1948 ('Exclusive Report on the Tristan da Cunha Fishing Industry Expedition').

Sphere, The, 11 Aug. 1945 ('Three Naval War Secrets Recently Revealed'); 29 Jun. 1946 ('A New Era for Tristan da Cunha').

Tablet, The, 22 Apr. 1933 ('The Colonial Secretary's Decision'); 31 Mar. 1934 ('The Lonely Island [1]'); 7 Apr. 1934 ('The Lonely Island [2]'); 30 Nov. 1935 ('Tristan da Cunha').

TDC Newsletter (Tristan da Cunha Association), *passim*.

Tristan da Cunha Facts and Figures

Location	–	South Atlantic Ocean Latitude 37° 3′ South Longitude 12° 18′ West
Size	–	98km2 (38 square miles) Average diameter 12kms (7.5 miles) Coastline circumference about 40 km (25 miles)
Highest point	–	Queen Mary's Peak 2,060m (6,760ft)
Status	–	UK Overseas Territory with St Helena and Ascension Island
Capital	–	Edinburgh of the Seven Seas
Currency	–	Pound sterling
Population	–	244 (2020)
Language	–	English
Religion	–	Christian
Time	–	GMT
Electricity	–	240V, 50Hz. The standard electrical socket is the 13-amp flat or round pin used in the UK.
Postal address	–	Tristan da Cunha, TDCU 1ZZ, South Atlantic Ocean, (Via Cape Town, South Africa)
Website	–	www.tristandc.com - for all updated contact details

Acknowledgements from Richard Grundy

This book has taken many years to turn from concept into reality. When I was living in the friendly and generous Tristan community I learned about the islands from keen pupils, their families, expat colleagues and expert visitors. There were few books, and I always had the ambition of producing one that might combine the island's physical background with its history up to the present day.

I must start by thanking my housekeeper Martha and her husband Eric Glass who looked after me as a bachelor during my first term on Tristan and who inducted me into the social life of the village. Anne Green, teaching in the neighbouring classroom, was a huge help with Tristan Studies and we have kept in touch, especially during her second time as Chief Islander twenty years later. Through her husband Joe I got to know the mountain and was soon leading expeditions with pupils. Other trusted guides included Harold Green and Ernie Repetto, who took me to Nightingale aboard his beloved longboat Margaret Rose and taught me all I know about penguin guano! I was lucky to work alongside Education Officer John Cooper with whom I formed a close friendship and effective partnership that continues today as we are joint chairs of the Tristan da Cunha Association. Administrator Colin Redstone encouraged Tristan Studies, started my speaking career on a cruise ship, and gave me a hosting role with visitors. Links made with Michael Swales who led the Denstone Expedition to Inaccessible, and with Martin Holdgate who sent me his monograph, were established at this time. I left Tristan in 1985 with my head full of island life, and with a bulging file full of written records that helped later when writing this book.

In 2004 I met Administrator-elect Mike Hentley before he left the UK, and we formed a close partnership that resulted in expanding the TDCA Newsletter and starting the website. By then several of my ex-pupils and old friends occupied key roles and helped provide news and photos. These included Cynthia Green, Iris Green, Andy and Lorraine Repetto, Robin and Dawn Repetto, and Chief Islanders Ian Lavarello, Conrad Glass and James Glass. It has been such a pleasure to renew links with these capable people. Further Administrators provided reports and put me in touch with visiting experts. Sean Burns deserves special thanks for his help during two three-year terms and who remains a close confidant. My conservation education was led by colleagues based in South Africa: another John Cooper (SA) who set up the molly-ringing project, and Peter Ryan continues to be a great source of up-to-date knowledge. In recent years RSPB officers have been in the forefront of research and I specially thank Katrine Herian and Antje Steinfurth for their help.

The book plan was announced in the August 2009 newsletter and members invited to contribute. As a result, former Administrator Peter Wheeler made two trips to Somerset for extended interviews and loaned me his Tristan archives. I refused his offer of sponsorship as I couldn't promise a publication date and I regret that his death prevents him seeing the finished book.

Further significant contributions came from Jim Flint, who donated his excellent book and photo collection, Bob Shaw, who sent me his diary and images of the Royal Society Expedition, Stewart Miller who provided a report and pictures of the work of HMS *Leopard*, and Ole Magnus Mølbak Andersen who reported on the *Bornholm* voyage. A breakthrough came when Joan Umpleby discovered in an attic a collection of transparencies taken by her father Geoffrey Dominy, who had been the captain of the MV *Frances Repetto*. John Cooper (SA) arranged for this precious material to be digitalised by Ria Oliver of the Antarctic Legacy Project at Stellenbosch University and an example adorns the front cover of this book. Having met his father Roland during his last visit to Tristan, I am indebted to Torbjörn Svensson who recruited me to work aboard the MS *Island Sky*, thereby enabling me to return to the island. Torbjörn gave me images of Roland's paintings and a reprint of his 1960s book.

Webmaster Peter Millington has provided constant assistance through the book production process. Other key Association members who have helped include postal history experts Robin Taylor, Ron Burn and Mike Faulds, and Tristan Government UK Representatives Chris Bates and Chris Carnegy. Further help came from Sue Scott, Heather Broadbent, Jennifer Craig, and Islanders Ian Brown and Debbie Elsmore who loaned me her father Lars Repetto's immaculately handwritten diary covering 1962/63.

In 2019 I invited Neil Robson to join me to put the book together as I was impressed with the Tristan historical articles he had written for the newsletter. We agreed an action plan to work as co-authors and our resulting work has been a joint effort. Neil is almost entirely responsible for Part 2, but his influence extends to every chapter and his ideas permeate all aspects of the publication. Without Neil the book would still be on the drawing board, having missed another anniversary. I am immensely grateful for his huge effort, without thought of financial reward.

Association colleagues Michael Swales, Jim Kerr, Peter Millington, Chris Carnegy, Sir Martin Holdgate and John Cooper have helped with general editing, and Anna Hicks with aspects of vulcanology, but the book should not be regarded as representing the views of the Association, but rather those of the authors. Circumstances have meant that the text has been edited without island help, but most of the modern content has already been published in the newsletter or on the website. All images appear here in good faith as contributions for Tristan Association use, although the provenance of some of the photographs cannot be readily verified. I also thank regular newsletter printer Murray Wallace from Direct Offset in Glastonbury for designing the book and to printers CPI.

Finally, I pay tribute to my wife Margaret who gave up her own teaching career to join me on Tristan at short notice (we met and married in less than a month). She has tolerated, nurtured, and been a huge support of my various island projects ever since.

Index

D

E

F

G

H

I

J

K

L

M

N

O

P

Q

R

S

T

U

V

W

Y

Z